The Uranium Investment Book

By Michael H. Caldwell

Creative Classics Inc.

Kelowna, BC

Canada

ISBN: 978 – 0 – 9784620 – 1 – 7

Caldwell, Michael H., 1946 –

The Uranium Investment Book

Printed by: Friesen Printing Corporation

Graphic Artist: Donna Szelest

Introduction by: Sidney Himmel

Research: Peter Mitham, Toby Osborne, Stan Sauerwein, Tony Wanless

PRINTED IN CANADA

TABLE OF CONTENTS

INTRODUCTION

A renewed interest in nuclear energy as a non-polluting source of power means that throughout the world uranium is increasingly becoming the most important source of electrical energy. A metal mined from the ground, uranium in its purified form, called U3O8, is a yellow powder: hence its other common name, yellowcake. Yellowcake is converted to Uranium Oxide for use as uranium pellets in the fuel rods of nuclear reactors, which use fission of atomic nuclei in nuclear chain reactions that releases electromagnetic energy, which in turn powers steam turbines to produce electricity. In North America, nuclear reactors account for 20% of electricity production, while carbon-burning "dirty" methods such as coal, natural gas and oil account for 72%. Clean hydroelectricity accounts for 8%. But the pressure is on to change those proportions, and as the global demand for nuclear energy increases, so will the demand for uranium, and therefore the price of uranium.

The advantages of nuclear energy are clear. Uranium consumed in electricity production has low cash cost compared to costs of other fuels. Per kilowatt hour the cost of nuclear energy is 1.7 cents; coal is 2.2 cents, oil is 8.0 cents, and gas is 7.5 cents. One ton of uranium generates as much power as 25,000 tons of coal. Uranium makes up only 5 per cent of the costs of operating a power plant, while natural gas approximates 75 % of plant operation costs. Nuclear power production is also proven and reliable technology. Other proven technologies that can produce dependable energy in quantity are the carbon-based organic fuels such as oil, gas, coal, and hydro power which results from harnessing water flow in natural falls or dams. New technologies such as hydrogen, wind, solar, geothermal and ethanol are unproven at the present time, either in terms of technology, cost, or the ability to produce sufficient amounts of energy. Also, with technological advances, nuclear energy is clean energy, producing only small amounts of waste.

This nuclear renaissance elevated the price of uranium from US$36 per pound in 2006 to US$80 per pound (in the fall of 2007). This surge was followed by a downward correction during the following summer, but the correction appears to be complete and a new bull market to have begun in October. Higher uranium prices will produce increased profits for uranium miners and significant access to equity funding for well-managed uranium exploration and development companies.

Demand for uranium to power nuclear reactors has increased for several reasons. Concerns about global warming caused by fossil fuel burning combined wth the very high prices of oil and natural gas has moved public favor toward nuclear power in many countries. During the early 2000's nuclear reactor utilization increased dramatically to over 100% capacity and the number of planned new reactors increased dramatically. There are currently 439 operable nuclear plants in the world. A further 34 plants are under construction, 81 additional plants are planned, and an estimated 225 plants are currently proposed. By 2030 it is estimated that the demand for nuclear power could rise from 370 Gigawatts per year to 700 Gigawatts per year. Estimates are that over the next 25 years the demand for uranium will increase from the current 180 million pounds per year to 350 million pounds.

At the same time, uranium supplies have dwindled. For much of the last 25 years world utilities relied on low-cost uranium inventories resulting from large build out of nuclear reactors during the 1960s and 1970s. Also, the highly enriched uranium supply from Russian warheads entered world uranium markets after the fall of the communist regime in 1990. These inventories were consumed between 1990 and 2005, as anticipated uranium demand from China, India and South America exploded, ending the era of excess uranium supply.

Meanwhile, two major changes have occurred in the uranium market. Hedge funds now participate in the uranium spot market (currently holding an estimated 5% of uranium production), increasing the volatility of uranium prices and causing the spot price of uranium to respond more quickly to changes in professional investor thinking. At the same time, uranium futures trading was introduced on the NYMEX exchange in May 2007, bringing professional financial investors into the investing mix along with classic buyers and sellers – utilities, mining companies and industry traders. This reinforced the sensitivity of uranium prices to general economic trends.

The world market capitalization of public uranium companies is tiny compared to that of public oil and gas and other resource companies. This alone provides the possibility of considerable upside in uranium public company shares. There are perhaps 300 or more public traded companies that have uranium in various stages of exploration, development and production. Investment in these companies requires careful analysis and a detailed understanding of the industry. Several key investment criteria exist. A company's ability to deal with volatile commodity prices is important since, for junior companies, financing risk is always an important factor that affects stock price volatility. Many of the potential new uranium deposits are in countries with significant geopolitical risks. One must consider environmental risks and public safety risks, as well as government policies with respect to these risks. Further, different projects will have different levels of technical risks, which must also be analyzed. A key investment criterion at any time is the quality of management. Quality management requires excellent technical skills, outstanding financial capabilities, and top managerial skills.

As of the time of writing of this introduction in late October, 2007, the spot uranium price is firming up again and so are the NYMEX uranium futures prices. Russia plans to build at least 2 reactors each year through to 2025. China plans to reach 40 Gigawatts of nuclear energy by 2020 – about 40 new reactors. India plans more than 25 new reactors by 2020. A mutual fund that holds uranium as its only asset, Uranium Participation Units ("U" on the Toronto Stock Exchange) is seeing ongoing unit price increases. Excitement has re-entered uranium equities as well.

Welcome to the great Uranium Bull Market.

Sidney Himmel, CEO
Trigon Uranium Corp.

Adriana Resources Inc.

The office of Adriana Resources is well appointed, although not ostentatious. Ensconced on the 18th floor of the Vancouver office tower that also houses the B.C. Securities Commission, the office radiates an aura of quiet professionalism. Looking down from its wide windows to the streetscape below, a visitor catches a glimpse of the busy blue harbour a few blocks away, and just below at the street level sees children playing on a patch of grass in front of the Vancouver Art Gallery, once the city's main courthouse. The impromptu playground is nestled among other glassed office towers reflecting the noon-day sun, comprising a representative tableau of the city as it exists today amid a financial boom, driven in no small part by mining and other resource related industries. Like the industries that are fuelling this boom, the office district manages to balance modern engineering with green thinking. Trees and grass surround each of the buildings, reflecting the combination of nature and business that is Vancouver today.

The office is a comforting combination of optimism and practicality that isn't lost on Adriana CEO, President, and Director Michael Beley. Beley, a mining veteran, opened his first office in the area more than 30 years ago. "When we opened our first office it almost killed us and we sat around wondering how we were going to pay for it," Beley recalls with a fond chuckle. "But we had talked to some older miners and they said we should have an office that reflects our values, so we stuck with it. We wanted investors and partners to understand that we were solid, that we weren't fly-by-night. We were going to be in this for the long haul and were out to build lasting relationships."

Clearly Beley has done that. By we, he means he and his long time business partner, CFO and Director of Adriana, Richard Barclay. The two men have been building mining companies together for 37 years, often with other business partners they have formed relationships with since they opened that first office. Beley and Barclay met in a tent on Vancouver Island when both were young geologists and prospectors. The two hit it off immediately and have remained partners because they share the same philosophies regarding how a mining business should operate. The partnership has been helped by the fact that the two are opposite personalities. "We're Mutt and Jeff," Barclay says, "I'm the introvert, the financial guy, and Mike's the extrovert, the marketing guy."

The values the two held from the beginning continue to drive their business decisions today, although they now are supplemented by something even more valuable in the sometimes-wild world of mining exploration and development—the experience that comes with developing several successful mining ventures. In 1982, Beley and Barclay co-founded Bema Gold Corporation and assembled the management that built it into a $3.5 billion company which merged with Kinross Gold Corporation in 2007. They also co-founded Eldorado Gold Corporation in 1992, which became a $2 billion company, and worked together at Nevada Pacific Gold Ltd. which ultimately merged with U.S. Gold Corporation in 2007. Beley was also co-founder of Polaris Minerals Corporation in 2002. Both men were partners in Bema Industries Ltd., a private exploration contracting and geological consulting firm with offices covering western North America and which managed Bema Gold Corporation. During their tenure at Eldorado, the company developed two open pit gold mines in Mexico; defeated an attempted hostile takeover; gained control of an Australian junior gold producer; participated in the strategy leading to a share-asset swap with South Africa based Gencor Ltd.; amalgamated with HRC Development Corporation; and participated in the management of the São Bento Gold Mine in Brazil and the development of significant gold assets in Turkey.

Beley and Barclay's long partnership has remained intact and been successful because they share the same beliefs; that experience, values, and long-term relationships are the bedrock of a successful mining venture. Beley explains. "It takes a lot to be successful in this business because you have to balance a lot of things. So experience really helps. This is especially true when it comes to attracting investment. If you're an investor and you're at a mining convention with 1200 exhibitors, you're going to have to target your information gathering. Isn't it logical then to go with someone who's been there, and come back? Doesn't that give you a feeling that there's a better chance of success?"

The partners' latest venture is almost a carbon copy of their previous projects, although with Adriana Resources Inc. (TSX-V: ADI) they're branching out beyond their earlier focus on gold. Their vision is that Adriana, founded in 2004, will ride a long up-cycle in commodity and mineral prices to become a world class mining company—a mid-tier multi-commodity mining producer—that is well diversified globally and contains a balanced collection of quality production assets. The company plans to foster this corporate growth by maintaining a sustainable pipeline of strategic exploration and development properties. "You can't rest on your laurels in this business," Beley says.

"You still have to produce, to walk your talk. Our belief is that the drill to success ratio is two percent, so we always ask what we can do to change the odds. Because of our management experience we believe we can change the odds—because we have access to money, to management, and to a great team. For example, we recently created a new company called Hawthorne Gold Corp., which acquired a well known gold project, and built a team to go to market. The IPO was done at $0.65, on the first day of trading the stock opened at $1.20 and traded up to $1.50. Investors piled in because of the project's proven management."

These values and experience manifest in how the partners approach mining ventures today, an adventure that can sometimes appear as a very complex puzzle. They understand that development of a producing mine includes many variables and so form a detailed strategy for all aspects of a venture. Against a management backdrop of climbing exploration and development costs, ever-increasing investor knowledge and information requirements, environmental considerations, and a need to become involved with people indigenous to the region in which they are developing mines, mine development often involves intense collaboration. Costs must be contained, environmental impacts must be assessed and addressed, and workers must be found. Clearly, no explorer can handle these myriad tasks singly, so the partners collaborate as much as possible.

There are four key elements to Adriana's strategy: to maintain a diversified portfolio of properties, to form strategic partnerships, to create effective working relationships with local communities, and to maintain a skilled and experienced management team. Adriana is committed to maintaining a balance of late-stage projects that are being advanced to production and key exploration targets with excellent discovery potential. Advancing late stage projects builds shareholder wealth through sustainable development, continued growth of mineable reserves, and the development of a strong financial profile through sustained cash-flow. In its choice of exploration properties, Adriana focuses on identifying other advanced projects that match the corporate growth objective of building a well-balanced mineral asset base that is strategically diversified globally.

In terms of partnerships, Adriana strives to form strategic relationships with other companies in all aspects of the mine operation cycle, from production requirements to product delivery mechanisms, which will create increased efficiencies and cost savings. The company is also committed to strategically positioning itself in countries with a mining friendly climate, available skilled workforce and a progressive and developed infrastructure program. Adriana ensures effective working relationships wherever it operates by fostering the growth and success of the communities where it operates, acting as leaders in environmental and social responsibility, and striving to create a safe and rewarding workplace. And finally, the management team that Adriana brings to all its projects is built to reflect the company's entrepreneurial spirit, integrity, passion, trust, and hard work.

Adriana's successes over the past few years are a testament to these strategies. Since 2004 the company has made several important acquisitions, including the MIE nickel-copper-PGE Project in Nunavut. This 630-square-kilometre property is part of an extensive magma conduit in northern Canada. The MIE Project draws its name from one of the world's largest continental-type magmatic events, the Mackenzie Igneous Event. This event deposited a huge volume of magma—estimated at 5 to 10 million cubic kilometres—across northern Canada. Rare geological events such as these are widely known to host large, rich, platinum group element (PGE) and copper–nickel deposits. To date, Adriana has completed extensive geological and geophysical exploration programs at MIE and commenced a diamond drilling program with significant results.

Another exciting acquisition is the option to purchase the Lac Otelnuk Iron Project property, located 170 kilometres north of the town of Schefferville, Quebec. This town once served as a centre for the iron mining and processing operations of the Iron Ore Company of Canada. The property lies within the Labrador Trough, one of the largest iron ore belts in the world. The belt contains world-class iron deposits that have been continuously mined since 1954. Based on systematic and well-documented past exploration, and on its own initial evaluation of the property, Adriana believes that a multi-billion-ton iron resource lies within the 45–kilometre iron formation belt contained within the property. Historic concentrate grades and weight recoveries produced from metallurgical tests of surface samples, drill core, and a bulk sample conducted by previous operators compare favourably with active iron-ore operations in the region. To date, Adriana has initiated the first phase of the work program at the Lac Otelnuk property, completed construction of an exploration camp, and commenced a drilling program to bring the reporting of the existing multi-billion tonne iron resource into compliance with current National Instrument 43-101 disclosure standards.

Adriana has also acquired an option on the Mustavaara Mine, a past producing iron-titanium-vanadium mine in Finland. The property is located in central Finland and consists of four exploration claims totalling 356 hectares that are accessible by paved road. The deposit was discovered in 1957 and the mine, which in 1978 accounted for some ten percent of the global supply of vanadium, was in operation during the 1970s and 1980s. The mine closed in 1986 due to low metal prices. However, a consultant's study showed the mine still contains vanadium content in magnetite at a concentration of 0.91% to 1.65% V2O5, a high grade compared to other vanadium mines currently in production. The reported remaining Measured Mineral Resource at Mustavaara is 30 million tonnes at an average grade of approximately 16.8% magnetite with a 0.91% vanadium content. The Mineral Resource is based on a cut-off grade of nearly 11.9% magnetite with a 0.75% vanadium content. So far on this project, Adriana has completed an NI 43-101 technical report and a bulk sample. The company has also engaged Behre Dolbear International Ltd. to begin a scoping study.

One of the newest projects for Adriana is the Bear Valley Uranium Project (BVU) in Nunavut, Canada. The BVU Project covers some 750 square kilometres along the eastern edge of the Hornby Bay Basin. The Hornby Bay Basin represents an under-explored sandstone basin with significant uranium potential that shares many geological characteristics with the nearby Athabasca and Thelon Basins. Numerous high-grade unconformity style uranium deposits have been identified within the Athabasca Basin, and mining of these deposits currently generates approximately one third of the world's uranium production. The Hornby Bay Basin was a focus of uranium exploration during the 1960s to the early 1980s. Triex/Pitchstone's Mountain Lake uranium deposit that occurs within the Dismal Lakes Group sediments was discovered during that period. With the rise in uranium prices, there is a renewed interest in the uranium exploration activity in the Hornby Bay Basin.

The BVU project is something of a departure for the partners, who have spent most of their lives finding and developing gold mines. But it became a natural extension of their move into general mineral exploration, and emerged as a result of that mining thrust. They were exploring the adjacent MIE project in 2004 when a prospector, Gordon Addie (Adriana's current VP Exploration), noted the uranium potential and similarities to the Mountain Lake uranium deposit. "Uranium for us was an accident," Beley explains. "Uranium was not first in our minds, and it was not our focus to go there.

Accident or not, once they looked at the uranium exploration proposition, Beley and Barclay could see a very tangible upside to uranium exploration. Prices have been climbing since the low point in 2002 when the world had little use for uranium and it was languishing at $10 a pound. As the increased demand for nuclear power in Europe and Asia combined with a shortage of oil and an expected shortage of uranium supply, prices began an upward trend that reached $135 a pound in 2007. They're predicted to sustain those heights and perhaps ascend higher because demand for Uranium U3O8, the active ingredient of uranium ore that powers nuclear reactors is increasing, and it could be a decade before new mines go into production. At the same time, one of the world's biggest uranium producers was exploring nearby, and benchmarked it as a top world wide uranium target area, validating the partners' initial information that the area was a promising uranium field. Further, investors had been watching the uranium price climb and were starting to flock to it. Uranium was a sexy mineral again, and having your hand in producing it would be good from an investor relations point of view. A uranium property could help fuel the kind of investor interest that would help finance exploration for more mundane metals. As experienced mining company builders, Beley and Barclay could see that uranium exploration made good sense.

In 2005, Adriana carried out a geophysical survey of the area, delineating a stratabound conductive horizon in the sedimentary rocks of the Dismal Lakes Group within the Hornby Bay Basin. The anomalous zone (named Alpha Horizon) is approximately 16 by 6 kilometres in extent. It is near surface and nearly horizontal. It is also along strike of structures known to contain uranium occurrences. A more detailed survey of the area was recently completed and modeling of this survey is ongoing. The company's 2006-07 Bear Valley uranium exploration program focused on prospecting for stratabound and structurally controlled uranium mineralization in the Dismal Lakes Group sedimentary rocks along the north-eastern margin of the Hornby Bay Basin. Uranium mineralization in the Hornby Bay Basin is known to occur in and adjacent to structures within sediments of both the Hornby Bay and Dismal Lakes Group sediments.

As a result of the 2006 uranium prospecting program, numerous surface uranium anomalies in the Hornby Bay Basin were discovered along faults and fractures near the basal unconformity and in overlying stratigraphic horizons. Stratabound uranium occurrences were also found associated with the graphitic horizons in the basement rocks beneath the sedimentary rock package. Highlights from the program include a subcrop sample collected from the Tabb Lake area that returned 7.281% U308, confirming anomalous uranium mineralization reported by previous operators. The 2005–2007 exploration programs have led Adriana to consider the All Night Lake Area, Pointer Lake Area, and Tabb Lake Area as promising for further exploration. Adriana is planning a diamond drilling program on its 100% owned property in 2008.

In 2006, Adriana announced the formation of the UNAD Uranium Joint Venture with UNOR Inc. This 50/50 joint venture consists of twenty-nine claims that cover approximately 39,284 hectares. The properties are located on the Eastern edge of the Hornby Bay Basin in Nunavut, and adjoin Adriana's Bear Valley and MIE projects, and UNOR's Coppermine River Property. Five of the claims are located in the Kendall River area of the Hornby Bay Basin. These claims include several historic uranium occurrences cited in assessment reports, including: Munch Lake, Bear Valley, and Tabb Lake. Adriana's joint venture with UNOR Inc. has improved the effectiveness of both companies by allowing the sharing of knowledge and resources. Cameco, a world leader in uranium exploration, has also recently entered the Hornby Bay Basin: through a 19.5% investment in UNOR Inc., and a large-scale staking program, which was then immediately followed by an airborne geophysical survey. Adriana has spent approximately $350,000 of the spring and summer 2007 exploration budget for BVU and MIE projects towards its obligation under the UNAD Joint Venture with UNOR.

The alliance with UNOR was typical of the process Beley and Barclay have established over their long history of building mining companies. Bema Gold was formed out of a collaborative relationship of several companies, and Eldorado also involved a collaboration of sorts between the mining company and the Mexican communities in which it was involved. "The objective in this case is information gathering," Beley points out. "Uranium involves a long development cycle, and we were using similar techniques on ancillary properties. So we went to them and said let's stake together and share information about the properties we're both next to, and that will tell us more about the properties each of us owns. It fits with our general strategy to diversify as much as possible— in minerals, in costs, and in reaping the rewards. In a sense, it's back to our value systems again, to work together with people in a trust relationship."

The alliance is also a creative method that allows Adriana to begin to address an impediment faced by all exploration and mining companies today—the lack of trained workers. During the long period when metal prices were in the doldrums, much of the mining infrastructure crumbled, and workers deserted the industry for more stable jobs. Training of new workers dwindled and interest in mining among younger people slumped. As a result many of those workers that remained in the industry are nearing retirement, or are no longer willing to sacrifice their quality of life for their jobs. As with many other industries, human resources today has become as big a problem as raising capital was yesterday. "In the old days, I would be in the field for four months at a time," Barclay recalls. "Now, the standard is four weeks in, two weeks out. That means more workers are needed, but they're harder to find. This year we are really being impacted by the lack of staff. For example, there's a surplus of drill rigs around but there's no one to run them. One drilling company we are aware of owns 27 rigs under contract, but 5 are idle as they have no available personnel. There's nothing that can be done about it, you just have to stay the course. We're thinking long-term anyway."

In fact, long-term thinking is one of several aspects of a plan the partners have devised to deal with the ongoing staff shortages faced by everyone in the industry. By thinking over a multi-year period, perhaps a decade long, the Adriana partners created a framework to fill their worker pipeline in future when the need for them will be even more crucial. The framework might mean the venture will take longer, cost more, and feature increased risk, but it will also go farther to guarantee its success. This framework includes building a nucleus team around which other workers can be deployed. Because it has maintained a team for years, Adriana is well on the way to achieving this goal.

Another of Adriana's strategies involves hiring multi-nationally. Like other forward thinking mining companies, Adriana is scouring the earth for skilled workers it can bring in to help with its exploration and development. In some cases these workers might be trained in other areas and will need to adapt. "But," says Barclay, "that can be done if you're resourceful in finding them." In addition, Adriana sees opportunities in providing training and advancement opportunities for people already in the company. "You can take people from further down the chain and move them up. The shift involves taking more responsibility and isn't for everybody," Barclay adds. "But it's a good way to maximize resources and reward employees at the same time."

The company also believes that the human resources crisis can be lessened by treating indigenous people as a resource. At the Hornby Bay Basin/MIE projects, Adriana is working closely with the local Inuit communities to train people in mineral exploration and mining procedures. "It's naturally logical to use local people as human resources," Barclay says, "because they know the terrain. Also, they have a vested interest in making the venture a success in terms of sustainable economic benefits." Working closely with indigenous people who live in the exploration area is also part of Adriana's strategy to commit to the development of environmental and social prosperity in the regions where it conducts operations. Today, working in harmony with local communities and Aboriginal groups is extremely important to achieve success in the mining business, and miners as experienced as Beley and Barclay clearly understand that. The Hornby Bay Basin uranium project lies within the new Canadian territory of Nunavut, where land claims were settled in the 1990s. In Nunavut, therefore, land claims do not create an impediment to mining as they sometimes do in other parts of Canada where they remain unsettled. In fact, the land claims settlements in Nunavut provided a structure for how companies should deal with the local people, the Nunavummiut, who make up part of the population of Inuit people who primarily populate Canada's north. Under this structure, Adriana deals directly with the Nunavut government, which is looking to mining—the territory is also seeing increased activity in gold and diamond mining—as an economic driver to improve the standard of living for the region's population of just under 30,000 people.

While Adriana has launched a program to train Inuit for work in the western Nunavut exploration area—one of the most remote places in the world just below the Arctic Ocean and southwest of Victoria Island—it is also taking pains to help preserve the Inuit culture, which has come under attack in recent years through increased interaction with southern Canada. For example in ancient Kugluktuk, where the Coppermine River meets the Arctic Ocean, Inuit people built a culture around copper artefacts. However the art of copper working has been lost, and so Adriana is trying to revive the part of the Inuit culture that revolved around it. It is doing so in part by funding the reintroduction of traditional copper related skills and artistry to encourage creativity and learning among young Inuit. "We have a good relationship with the Nunavut government because we're quite active in the area with the MIE and Hornby Basin projects," Beley explains. "It's in our and their best interests to revive the culture there, but in a 21st Century setting. The future for them isn't just about mining and industry. It's also about bringing the best of the past culture along with them into the future."

Part of working with indigenous populations also involves environmental stewardship of the land being explored and mined. The days when miners would arrive in an area, produce ore until it was played out and then leave behind giant tailings ponds, open pits, and other scars on the landscape, are long gone. Today, mining is conducted with extreme care regarding environmental impact. This requires extra planning, but is essential in a climate as harsh as the northern Arctic, where environmental impacts can be very long lasting. For example, Adriana flies all its garbage out of the area to Yellowknife, the capital of the neighbouring Northwest Territories. "It's not something you think about," Beley says of environmental care and remediation. "You just do everything you have to do to preserve the environment. It's more than just providing jobs and training and a future for people. We also have to ensure the area they live in remains the same. It's part of their life."

Beley and Barclay have parlayed long experience working with indigenous people in Canada, Mexico, and Finland to stack a strong hand that increases the odds of success for their current Adriana projects. But experience with people does not simply involve the residents of regions in which they work. They also recognize that to be successful today, a mining company must also have experienced people internally who can work together to help improve the chances of success. Today, exploration for any metal, whether it be gold, iron or uranium, and then development of a working mine requires a team of experts in their own fields who can swing into action when their skills are required. To that end, Beley and Barclay have assembled a management team at Adriana that is almost as experienced as they are.

Adriana's Director of Business Development Frank Condon, BASc, PEng, has over forty years of experience as a geological engineer in the mineral resource industry. His experience includes thirty-five years with Noranda Group Companies. During Frank's tenure with Noranda he was involved in mineral exploration, mineral property evaluation, mine development and mine operations. While based in the US for seventeen years, Condon held positions of District Geologist, Senior Evaluation Engineer, Project Manager, and General Manager of Noranda's phosphate mining operation. In 1987, Condon relocated to Noranda's head office in Toronto as Director of Business Development, where he was involved in evaluation and acquisition of advanced exploration and mining projects internationally. Since 1999, Condon has been a Consultant and is currently a Director of Adriana Resources. Condon graduated from the University of British Columbia with a Bachelor of Applied Science Degree in Geological Engineering. He is a member of the Professional Engineers of Ontario and the Canadian Institute of Mining and Metallurgy. Mr. Condon is the recipient of the distinguished CIM Members Award for 2003 in recognition of his unique career and contribution to the mining industry.

Gordon Addie, BSc, has been the Vice President of Exploration for Adriana Resources Inc. since July 2005. Addie is also the Vice President of Hawthorne Gold Corporation, a TSX-V listed Canadian-based gold exploration and development company, and the President & Director of 5050 Nunavut Limited, a wholly-owned subsidiary of Adriana. Addie earned his Bachelor of Science degree in Geology in 1986 from the University of British Columbia and has over 20 years of exploration and mine geology experience.

Adriana's Controller since July 2006 is Patrick McGrath, BComm, CGA. McGrath is also currently the CFO and Secretary for Hawthorne Gold Corp. McGrath earned his Bachelor of Commerce degree from Memorial University of Newfoundland in 1995 and is a Certified General Accountant. Between December 2004 and January 2007, McGrath was CFO of Northern Sun Exploration Company Inc., a TSX-V listed oil and gas exploration company with properties in western Canada.

Anthony Kovacs, Manager Exploration, has over ten years of experience applying geophysics to mineral exploration challenges. Kovacs has previously worked on the Muskox intrusion (MIE Project) for eight years, perfecting geophysical surveys for a variety of mineral deposit models unique to this large igneous province. Before joining Adriana in 2006, Mr. Kovacs worked under contract with Anglo American as Project Geophysicist on nickel projects in Quebec, Manitoba, Alaska and Nunavut and iron oxide copper gold (IOCG) projects in Canada and Mexico.

Joe Griebel, MSc, PhD, Director and President of Adriana's subsidiary, Adriana Mineração Ltda., is the latest addition to the Adriana team. Griebel comes to Adriana after spending the past six years as President of Inco Brazil Limitada where he remained until the recent takeover of Inco by Companhia Vale do Rio Doce (CVRD). Griebel holds an MSc in geology and a PhD in mineralogy and has over thirty-five years of experience in international mineral project management. Eighteen of those years were spent in Brazil. Griebel is associated with Novamina Empreendimentos Limitada, an exploration services company based in Belo Horizonte, Minas Gerais, Brazil. Previous to his position with Inco, Griebel spent three years (1996–1998) as the Senior Vice-President of Exploration for Eldorado Gold Corporation, managing all of Eldorado's international exploration activities. From 1990–1996 he was the Director of Exploration for Unamgen Mineração e Metalurgia S. A., a subsidiary of Gencor South Africa, and a Director of Gencor's São Bento Gold Mine in Brazil. Griebel is responsible for all of Adriana's activities in the country of Brazil.

The advisory team at Adriana are just as important as the management team, and equally qualified. Michael Redfearn, BSc, PEng, is currently the Vice President of Operations for Hawthorne Gold Corp. Redfearn has over thirty-five years of mining, metallurgical, environmental, and construction experience and brings to the firm long time management expertise in the area of arctic and high altitude remote minesite operation. Prior to joining Hawthorne, Redfearn was the Vice President of Operations for bcMetals Corporation, a TSX-V listed mineral exploration and development company with assets in north-western British Columbia. He has also served as Mine Manager at the Cantung Mine in the Northwest Territories and with various Cominco Limited operations from 1986-2001.

John-Mark Staude, MsC, PhD, has over twenty years of diverse mining and exploration experience in precious and base metals. He earned a Masters of Science from Harvard University in 1989 and a Ph.D. in economic geology from the University of Arizona in 1995. He has held positions of increasing responsibility with Kennecott, BHP-Billiton, and Exploration Business Development with Teck Cominco as well as working with commodity-focused companies Magma Copper Company and several private companies. He is founder and President of Riverside Resources Inc. and has consulted to private investment groups. Staude has extensive mineral resource experience on four continents including participation in the discovery and district exploration at Mulatos gold mine in Sonora, Mexico and the Agua Rica copper–gold deposit in Argentina. He has been part of gold discoveries in Peru and Turkey, and has formed and led exploration companies in Romania, China, Turkey, and Bolivia. Staude has consistently added resources in known districts and helped in converting grass roots discoveries into new mining operations.

Alberta Star Development Corp.

Uranium was first discovered in the Great Bear Lake area of Canada's Northwest Territories in 1929, when adventurous geologist Gilbert Labine uncovered High grade silver–pitchblende veins at Port Radium. Other discovered veins at Port Radium, Eldorado, and Contact Lake were mined until 1940. One of the world's first uranium finds, the Eldorado mine was reopened a year later to supply five tons of uranium oxide to the United States Government for the Manhattan Project, which resulted in the development of the world's first atomic bomb. However, after the war, the price of uranium dropped and all mining and exploration in the area ceased in 1960.

After its brush with fame, the Great Bear Lake area slept, in mining terms, for more than 40 years, not only because the low market price of uranium all but killed exploration for the metal, but also because growing strength among the First Nations community in the Northwest Territories—the Sahtu Dene people—resulted in an anti-mining sentiment in the territory. In fact, mining companies for the most part have completely shunned this mineral rich territory because this perception made permitting extremely difficult.

The Sahtu's worldview was different from that of most exploration companies. Like many First Nations people, they felt they were only stewards of the vast landscape in which they lived, not masters of it. As hunters, their living rhythms were dominated by wildlife, particularly the caribou that migrated annually across the NWT's Barrens and south through the Great Bear Lake regions. Above all things, the Sahtu did not want to interfere with these migratory routes involved in the great annual migrations, which accounted for the majority of their food supply. Also, English is not commonly used among the Sahtu (Slavex), who have an entirely different culture than the Inuit to the North and the English and French cultures to the south. Even the Sahtu language is different, given to the use of visionary analogies and metaphors to explain concepts instead of the logical approach used by most common languages.

In 1998 the Sahtu Dene concluded land claims negotiations with the Federal Government and received title to vast sections of land in the Northwest Territories. Over the next four years they began to build a governing infrastructure that would allow permitting for mining exploration, but because the system was so new and fragile, most mining companies continued to ignore the region. In 2004, however, Alberta Star Energy Corp. President and CEO Tim Coupland activated an aggressive exploration strategy that relied heavily on the help from the Sahtu Dene. With more than 20 years of business experience with both public and private companies in both debt and equity financings, Mr. Coupland has wide expertise in raising over $60 million of equity capital and assembling highly seasoned teams of professionals, consultants, financial consultants, and mineral exploration advisors with proven track records in mineral exploration. Combining this experience with experience working with many First Nations and Métis groups in the Northwest Territories, Mr. Coupland was the lead negotiator who secured the first impacts and benefits agreement with the Sahtu Dene on their traditional lands. He also oversaw the rigorous permitting process with the Northwest Territories regulatory authorities responsible for issuing land use permits, water licenses, and conducting environmental assessments.

Alberta Star's exploration strategy was based on the belief that the most likely place to find uranium was where it had proven to be in the past, and it hit early on the Great Bear Lake region as a probable candidate. After all, the area was among the first proven reserves of uranium in the world, and only stopped producing because technology at the time couldn't advance the mine any farther and a large, more easily accessible, discovery had been made in Saskatchewan's Athabasca Basin region. Since then, however, the Canadian government had been continually studying the area's mineral geology and putting its findings into the public realm. Alberta Star, a multi-mineral exploration company based in Vancouver that had primarily been searching for gold in Ontario since 1996, saw in the findings a natural opportunity to stake the area. But it had to bring the local Dene people on side. This wasn't that onerous a chore for Coupland, who had previously become familiar with the Sahtu Dene and liked their personal values, leadership and progressive business models.

Coupland recognized that any exploration in the area would have to be sensitive to Sahtu Dene concerns about wildlife, and would also have to provide some economic benefit to the region's population, which was quite wary of any agreement with an exploration company. Coupland immediately began building a relationship with the local Sahtu Dene and cultivated it by hiring local people to study the region's wildlife, help minimize environmental impact, and help build (and rebuild in one case) an exploration camp. For two years, Coupland steadily supported the

Sahtu Dene in their efforts to form a modern regional economy. Alberta Star helped a Dene-owned helicopter company get off the ground by contracting as its primary customer. It also contracted with a Dene-owned airline to be its primary transportation carrier in the area and built and upgraded an airstrip at its Glacier Lake camp (which was shared with the Federal Department of Indian and Northern Affairs).

Also it worked closely with the local Dene government in Deline at the other side of Great Bear Lake, and its subsidiary, the Deline Land Corporation. Deline provided Alberta Star with environmental and bear monitors, camp helpers, and logistical support. At any given time during exploration, Deline had three to four wild life monitors and camp support workers steadily employed with Alberta Star. The company also continued building relationships with the Northwest Territory government, the Deline Land Corp., and the Sahtu Dene leadership through training programs, local business development, local recruitment and employment, and continual discussion. In 2006, Alberta Star set up the Branson's and Glacier Lake Camps and spent more than $2.3 million with the Dene for camp operations, permitting and licenses, community creation, and business development.

In return, Alberta Star is the first exploration company in 75 years to successfully stake, acquire, and permit one entirely contiguous land package in the uranium and mineral rich region. The Great Bear magmatic zone, which includes the Eldorado-Echo Bay mineral belt, is recognized by geologists as one of the most prospective iron oxide, copper, gold, silver and uranium (IOCG+U) regions in northern Canada. Alberta Star's Eldorado and Contact Lake projects encompass 98,027 acres containing five previous producing mines. Large rock specimens assaying over 0.60% U3O8 have been found on the property. The average head grade of uranium ore mined over the life of the Eldorado mine was estimated at 0.75% U3O8. The average grade for producing uranium mines worldwide is 0.15% U3O8.

Located on the site of ancient collapsed volcanoes, the Great Bear magmatic zone is Canada's premier IOCG+U setting. A relatively new discovery, IOCG+U settings are generally found in such ancient volcanic regions and traditionally feature bulk targets of low grade uranium and other materials, which can be mined with new technology via open-pit techniques that lower costs. The best known in the world is the giant (2 billion tonne) Olympic Dam Mine in Australia, which features large scale deposits of iron oxide, copper, gold, silver, REE, and uranium. Olympic

Dam style volcanic hosted hydrothermal iron-oxide–copper–gold deposits are attractive targets for exploration and development due to their polymetallic nature, high unit value, and enormous size and grade potential.

The Great Bear region hosts two of Canada's known IOCG deposits, and has produced uranium, silver and copper in the past. It also shows mineral occurrences along 400 kilometres and indicates a variety of polymetallic IOCG mineralization (Cu, Bi, Co, Ag, Au, U, Ni, Zn, etc.) The IOCG family in the Great Bear Magmatic Zone consists of a strata bound hydrothermal breccia containing molybdenum, copper, gold, lead, zinc, and tungsten and a structure controlled breccia with replacement mineralization of copper, silver, gold, cobalt, and uranium.

The Eldorado and Contact Lake claim block now consists of 11 contiguous claims located 5 kilometres southeast of Port Radium on the east side of Great Bear Lake and 470 kilometres north of the city of Yellowknife. The area consists of 98,027.77 acres and is comprised of two distinct areas: Contact Lake North and Contact Lake South. The Eldorado & Contact Lake IOCG+U project areas include five past producing high-grade silver and uranium mines. In Contact Lake North, the Echo Bay Mine produced 23,779,178 ounces of silver, and 6,900 pounds of uranium, the Eldorado Mine produced 15 million pounds of uranium, and 8 million ounces of silver and the area also included the Cross Fault Lake Uranium mine. The average head grade for the Echo Bay mine was 66 ounces per ton silver and the average head grade for the Eldorado silver–uranium mine was 0.75 % uranium. In the Contact Lake South area, The Contact Lake Mine, the Bonanza, and the El Bonanza mines were all former producers of silver, copper and high grade uranium.

The Contact Lake claim block and surrounding area cover extensive alteration zones including large mineralized gossans that can be traced for over one kilometre in length and over 200 metres in width. The Contact Lake area is located in the Eldorado Mineral Belt, which has been under-explored and has lacked advanced 21st century exploration geophysics. The Contact Lake Mineral Belt is approximately 15 kilometres long and is the northern extension of the same mineral belt that hosts Fortune Minerals NICO Gold–Cobalt–Bismuth deposit and the Sue-Dianne IOCG deposit. The Eldorado Mineral Belt, which is situated in the Great Bear Magmatic Zone, has recently been recognized by geologists, as one of the most prospective iron oxide, copper, gold, silver, and uranium regions in northern Canada.

The property contains a historic uranium tailings resource estimated to be approximately 910,000 tons of uranium at an average head grade of 0.75% uranium, and approximately 800,000 tons of silver tailings, at an average head grade of 66 ounces per ton. Alberta Star appointed SGS Lakefield for metallurgical processing and testing of the uranium and silver tailings and has completed a detailed sampling program of the uranium–silver tailings and waste dumps. It has also begun preparation for a definition drill program of the un-mined areas and extensions of the Eldorado uranium–silver mine with the intent of evaluating a potential resource to support recommencement of commercial production at the mine sites.

At the Eldorado-Echo Bay target, Alberta Star intersected 5.0 metres of 0.22% U308. A nine-hole preliminary drill program completed in the fall of 2006 was designed to re-evaluate the economic potential of the former Eldorado-Echo Bay silver and uranium mines. The initial drill results from the first nine holes discovered a new zone of hydrothermal and structurally controlled polymetallic vein breccias that are enriched in uranium, silver, gold, copper, nickel, cobalt, lead, and zinc. The completion of the drilling confirmed additional widespread polymetallic and uranium mineralization. Seven drill holes intersected multiple zones of intensely altered and mineralized polymetallic breccias with disseminated and vein hosted mineralization. All drill cores were prepared, bagged, and sealed by the company's supervised personnel and were transported by plane to Acme Analytical Laboratories Ltd. in Yellowknife, NT, where they were crushed and pulped, and then transported to Acme's main laboratories in Vancouver, B.C. for assaying. Acme is a fully registered analytical lab for analysis by ICP-MS and ICP-FA techniques.

The Eldorado South uranium target consists of 16 contiguous claims located south of the Eldorado uranium mine on the east side of Great Bear Lake and covers 37,202.32 acres. The anomalous area of the Eldorado South uranium claims includes several large radiometric anomalies of up to 2.5 kilometres in length and suggests a potentially significant near-surface uranium target. These large uranium anomalies have never previously been explored or drill tested and were therefore an important focus of exploration by Alberta Star in 2007. This was the first high resolution, multi-parameter regional radiometric and magnetic survey ever conducted over the Eldorado and Contact Lake IOCG and Uranium Belt using newly developed geophysics technology. Several of the larger anomalies show uranium radiometric signatures of comparable or greater strength than the known zones of uranium mineralization already identified on Alberta Star's uranium properties.

The primary Eldorado South Anomaly was discovered as a result of the completion of a high resolution, multi-parameter regional radiometric and magnetic geophysical survey conducted in July, 2006. The Eldorado South Anomaly is over 3.5 kilometres in length and the expression suggests a potential near surface uranium target. The radiometric profiles show a clear, well defined anomaly with a marked correlation of strong thorium and potassium ratio patterns.

The South Echo Bay gossan and the Mag Hill VTEM anomaly, located at the southern end of the southeast arm of Echo Bay, are viewed by Alberta Star as having the potential to host bulk tonnage mineralization. Drill results were successful in intercepting significant silver mineralization beneath a large pyretic gossan. The silver is contained in pyrite–chalcopyrite veins and disseminations within a kilometre scale phyllic and potassic alteration halo. The alteration zone is peripheral to an extensive zone of intense magnetite–actinolite–apatite alteration that was intersected in all holes drilled in the Mag Hill Program.

The Mag Hill target is the site of the most extensive hydrothermal magnetite–actinolite–feldspar–apatite plus sulphide alteration along the Contact Lake Mineral Belt in the northern part of the Great Bear Magmatic Zone. The host rocks are alkali (sodic or potassic) and or actinolite–epidote altered andesites. The core zone comprises a pervasive alkali feldspar–scapolite–magnetite–actinolite–apatite hydrothermal assemblage. The alteration signature and zone at Mag Hill is similar to that of the Port Radium-Eldorado uranium mine area situated on Labine Point, NT. The Mag Hill discovery zone and target area is defined by predominately pyrite mineralization, which is present intermittently throughout the Mag Hill region and in an extensive gossan at the southeast end of Echo Bay, with minor amounts of visible malachite. In addition, uranium, cobalt, nickel, copper, silver, lead, and zinc enrichments are present locally as veins, veinlets, and disseminations that extend outward in the southern and eastern extensions of the Mag Hill area.

This region is being targeted for detailed uranium exploration due to its notable similarities to the Port Radium-Eldorado mine region with regard to hydrothermal alteration and mineralization, and the large size of the associated hydrothermal system, the 2-square-kilometre core zone of magnetite–actinolite–apatite alteration, the kilometre-scale surface gossans, and the distinct lenses of polymetallic sulphide mineralization. Alberta Star analyzed over 121 surface grab samples from the Mag Hill Grid over a 5 square kilometre area in the southern portion of the Contact Lake belt. Twenty-five samples collected were variably enriched in uranium, silver, copper, lead, zinc, nickel, and cobalt. The rock samples assayed to a maximum of 0.69% uranium, 68.0 g/ton silver, 4.98% copper, 0.40% lead, 0.49% zinc, 0.23% nickel, 0.58% cobalt, 0.4 g/ton gold and 0.10% bismuth.

At the Echo Bay gossan at Glacier Lake initial drill results discovered a new zone of hydrothermal and structurally controlled polymetallic vein breccias that are enriched in silver, copper, nickel, cobalt, lead, and zinc. The Echo Bay silver-uranium target is located on the north side of Echo Bay. The company has intersected 10.5 metres of 117.1 g/ton silver, 0.14% copper, 0.46% lead, 0.21% zinc, and 0.18% nickel. This interval also included a higher-grade interval of 5.0 metres of 237.9 g/ton silver, 0.24% copper, 0.45% lead, 0.20% zinc, 0.37% nickel, and 0.09% cobalt, which contained a very high-grade interval of 1.50 metres of 739.5 g/ton silver, 0.19% copper, 0.56% lead, 0.45% zinc, 1.20% nickel, and0.22% cobalt. Completion of the five drill holes confirms additional polymetallic mineralization at Echo Bay. Three of five drill holes intersected multiple zones of altered and highly mineralized polymetallic breccias with disseminated and vein hosted copper, silver, cobalt, nickel, lead and zinc mineralization.

In its explorations, Alberta Star has also discovered a new large zone of hydrothermal and structurally controlled polymetallic breccias known as the K2 Discovery Zone. The breccias are enriched in copper, gold, silver, and cobalt situated on the company's K2 target site. The K2 target is located on the Contact Lake property, on the south side of Echo Bay. Preliminary drilling has yielded significant polymetallic results from an eight-hole drill program. Drilling and fieldwork has confirmed widespread polymetallic mineralization at several areas at the K2 target. All four drill holes reporting from a summer 2006 eight-hole drill program intersected multiple zones of altered and highly mineralized breccias with disseminated and vein hosted copper–gold–silver–cobalt sulphide mineralization.

The K2 target area is a new discovery in which significant polymetallic copper, gold, silver, and cobalt mineralization was encountered in a preliminary drilling program along a 250-metre strike length to a vertical depth of 300 metres. The mineralized zone remains un-delineated and open at depth and along strike in both directions. The mineralized zone is part of a regionally extensive linear zone of phyllic, potassic, hematite and magnetite, and magnetite–actinolite alteration zones, often appearing at surface as phyllic–potassic gossans. The mineralization occurs within the same suite of volcano-plutonic rocks that host other polymetallic zones in the Eldorado and Contact Lake mineral belt, including the former El Bonanza Silver–Uranium and the Eldorado Uranium mines.

Another new discovery is the Mile Lake Breccia, which is a high grade polymetallic breccia that is rich in copper, molybdenum, lead, zinc, silver, and tungsten, and is situated on the company's Contact Lake property on the south side of Echo Bay. Preliminary drilling of the Mile Lake Breccia has yielded some significant polymetallic drill results from an eight-hole drill program. Drilling and field work has confirmed widespread polymetallic mineralization at several target areas at Mile Lake, in both exploration rock sampling and in drill core. The Mile Lake Breccia features polymetallic mineralization and alteration that is intermittently exposed for over 2 kilometres in strike length, within a regionally extensive laminated volcaniclastic tuff.

Alberta Star has also embarked on a program to examine uranium and silver tailings from the now-moribund mines in the area. In 2006, the company completed a preliminary sampling of the uranium-silver tailings and waste dumps at the Eldorado IOCG and uranium project. The tailings are from the former producing Eldorado uranium mine (1933-1960) and Echo Bay uranium, silver, copper mines (1962-1982). The company has completed a detailed sampling program of the uranium-silver mine tailings and waste dumps and as a result of the sampling program has begun preparation for a definition delineation drill program of the un-mined areas and extensions of the Eldorado uranium-silver mine to determine if there is a potential resource to support the recommencement of commercial production at the mine sites. Assay results are pending for copper, gold, silver, nickel, cobalt, and uranium.

These mines are reported to have milled approximately 2.2 million tons of high grade uranium–silver ores leaving behind approximately 1.7 million tons of uranium–silver tailings. The Eldorado uranium mine formerly mined and produced 15,000,000 pounds of uranium at a head grade of 0.75% U308 and 8 million ounces of silver, plus copper, nickel, lead at Eldorado and Port Radium commencing in 1933. The Echo Bay mine produced over 23 million ounces of silver at an average head grade of approximately 66 ounces per ton up until its closure in 1982. Approximately 910,000 tons of uranium–silver tailings are currently contained in the Radium Lake and Cobalt Channel areas and an additional 800,000 tons of silver tailings are stored in the McDonough Lake containment area. An estimated 170,000 tons of uranium tailings were placed in surface depressions and in the Silver Point area and the remaining 740,000 tons were placed in the Cobalt Channel area of the Great Bear Lake.

Drilling results are often of great concern to geologists, but investors tend to look at company management as well. In this case Alberta Star has put together a strong team to ensure it can execute on its plan to revitalize the Northwest Territories uranium and IOCG mining.

Director Stuart Rogers joined Alberta Star in March 2007. Rogers has been involved in the venture capital community since 1987 and is currently the President of West Oak Capital Group, Inc., a privately held investment banking firm specializing in the early-stage finance of technology projects through the junior capital markets in Canada and the US. He has served as a director of client companies listed on the TSX-Venture Exchange, the Toronto Exchange, NASDAQ small Capital Market, and NASD OTC Bulletin Board. Currently, Rogers acts as a director or officer of the following companies which are reporting issuers in Canada: MAX Resource Corp., Consolidated Global Cable Systems, Inc., Randsburg International Gold Corp., Mexivada Mining Corp., Prophecy Resource Corp. and IGC Resource Corp.

Robert Hall is a company Director and Director of Field Operations. Holding a Bachelor's degree in Education from the University of British Columbia, Hall has over 14 years of industry experience, nine of which were in a senior management capacity with the Keg Steakhouse Restaurants of Canada. His accomplishments as a senior manager included receiving the "Top Keg Restaurant Franchise of the Year" award exemplifying excellence in all fields including sales, teamwork, leadership, human resources management, administration, financial analysis, and the ability to follow the Keg Restaurants rigorous high standards of excellence.

Director Tracy A. Moore is the president of MCSI Consulting Group, a Vancouver, B.C.-based firm he founded in 1990 that specializes in corporate finance matters, strategic planning, and business planning services. Between 1976 and 1990, Moore worked for three international accounting firms in restructuring, consulting, and audit positions. In addition to his consulting practice, he has owned and operated a variety of businesses. He serves on boards of directors and advises boards on corporate finance matters, business planning issues, mergers, acquisitions, and divestitures. Moore received a Bachelor of Commerce in accounting and management information systems from the University of British Columbia in May 1976 and was admitted as a member of the Institute of Chartered Accountants in British Columbia in 1979.

Director Michael Bogin, who has more than 25 years of diversified experience in asset-based lending and raising financings related to accounts receivable, inventory, equipment and real estate assets, trade financing, special purpose capital, mezzanine, and equity transactions. Since Nov 2003, he has owned and operated North Brooklyn Capital, an investment banking entity in Toronto, which provides a range of financial solutions and consulting services for commercial enterprises and entrepreneurs in the public and private sectors, covering a variety of business segments. Bogin has acted as a senior manager of Laurentian Bank of Canada, Vice President of originations for G.G. Capital Canada, and Vice President of Accord Business Credit Inc.

Ann-Marie Cederholm, CGA, joined the company in 2004 and currently serves as Corporate Secretary & CFO. Ms. Cederholm graduated from the University of British Columbia with a Bachelor's degree in Education in 2006. She also graduated from Royal Roads University with a Bachelor of Commerce in entrepreneurial management in 1997, and then received her CGA designation in 2003.

While this group provides the infrastructure that creates a strong exploration company, it is the Sahtu Dene who truly round out the Alberta Star team. An Athapaskan-speaking group of Dene and Metis who have lived in the Great Bear Lake region for centuries, the 2000-strong Sahtu Dene travel and intimately know a wide landscape occupied by moose, caribou, Dall's sheep and other small game and fish that are critical to their way of life. The caribou is of prime importance to them and is revered in the Sahtu Dene culture as the "giver of life", which is why they were so concerned about mining exploration disrupting the ecosystem in which the caribou exist. The Sahtu Dene society that exists in this vast lonely land is a milieu of strong cultural and individual identities created through kinship, social interaction with neighbouring peoples, interaction within family structures and deep knowledge of their history, which is passed on orally from one generation to the next.

People like the Sahtu Dene who have survived in harsh lands by being close-knit do not easily allow in their midst strangers who may disrupt the fragile social system that has been built over time. It is a testament to the planning and hard work of Alberta Star that it painstakingly took care to recognize, honour, and include the isolated and ancient Sahtu Dene before attempting to explore for uranium that will supply energy to the world in the 21st Century.

Cash Minerals Ltd.

The Yukon Territory entered the world's imagination as a source of undreamed gold wealth in 1896, when Tagish Charlie, Skookum Jim Mason and Kate and George Carmack discovered rich placer gold deposits in Bonanza Creek while fishing in the nearby Klondike River. In the stampede that followed, one of the greatest gold rushes the world has ever seen, the Yukon entered the world economy as a premier mining district. Over the next four years, $US51 million worth of gold was discovered and produced in the area around Yukon Territory's Dawson City, the heart of the gold rush. This is equivalent to about $US1 billion today. Historians say approximately four thousand people—most of them experienced miners—accumulated great wealth during the Klondike gold rush.

Alongside these experienced miners were tens of thousands of fortune hunters, many of whom had never so much as seen gold and had never left their hometowns before they came to the Yukon. A few of them got lucky, but many did not. The majority lacked any knowledge of mining, and most of them were unprepared for the demanding work involved. Even more were unprepared for the harsh and unforgiving climate. Usually, following unimaginable financial and physical hardship, those who made it back home considered themselves lucky to be alive and to have a few pennies left in their pockets. Workers who staked this area rarely realized a return on their investment and in many cases they never saw their friends or family ever again.

Times have since changed, in both the Yukon and the mining industry, though some similarities still exist. Today, advances in mining technology have produced highly efficient and specialized equipment. The mining business is also more complex; materials mined today are substances that no one had heard of during the gold rush. However, two fundamental qualities play a big role in the mining industry just as much as they did in those days: experience and practicality. Today, uranium is fast becoming the metal that has fortune seekers excited, and once again there are a lot of players in the Yukon with far more enthusiasm than know how.

Cash Minerals, a Canadian-based energy company focused on uranium exploration, could be compared to the miner who headed for the Yukon after he had already worked in the gold rushes of the Fraser Valley and the Cariboo Region. This miner was not lured by the romantic stories of how a rank amateur could get rich by simply picking up gold nuggets from the ground. Instead, he was someone who knew how to apply his knowledge to get a better of understanding of the landscape, and how to talk to investors about realistic expectations based on focused efforts. He was someone who understood that, while there are great fortunes out there for the lucky few, being one of the lucky few requires knowledge, experience and a lot of hard work.

In 1985, Cash Minerals began operations in the Yukon, under the name Silverquest Resources Ltd., and focused primarily on coal exploration and development. In 1991, Silverquest Resources changed the company name to Cash Resources Ltd. and subsequently to Cash Minerals Ltd. in 2001. By 2005, the company was a diversified exploration company with interests in coal, uranium and alternative fuels. After successful results produced from uranium exploration programs conducted in 2005 and 2006, it was clear that Cash Minerals' Yukon uranium portfolio held great promise and warranted additional attention by way of an expanded exploration budget and a newly created and highly specialized geological team. Thus, in 2007, the company launched its largest exploration program to date, committing $17 million (Canadian dollars) to uranium exploration in the Yukon. This program would use six drills and employ a staff of over 100 people to assist the company in achieving a clearly defined objective: to define a uranium resource at one or more properties in the Wernecke Uranium District, Yukon Territory. In achieving this objective, Cash Minerals would also be putting a new uranium discovery on the map.

Some could say Basil Botha, President and Chief Executive Officer of Cash Minerals, has the mining business in his blood. Botha grew up on a gold mine in Rhodesia, the son of a seasoned miner. Botha began his mining career working for a German mining company and focused his interests on energy resources rather than gold. He has since worked in the mining industry in Indonesia, China, United Kingdom and South Africa, specializing in start-up mining operations, mergers, acquisitions and corporate finance. During these years, Botha developed a solid track record of building early-stage resource companies into primary producers. In 2001, Botha moved from involvement in management with Reef Coal Mining Ltd. in Johannesburg to a role as a Financial Analyst with Reef Asset Management in Vancouver, Canada. In 2005, Botha was appointed President and CEO of Cash Minerals.

While meeting with Botha at his offices in downtown Vancouver it becomes quickly apparent that he is a skilled expert; someone who has over 30 years of mining experience under his belt. Relaxed and confident, Botha projects an aura of practical professionalism. The Cash Minerals offices also reflect this practical attitude. The company is located on the 18th floor of the Oceanic Plaza in Vancouver's financial district. The décor is minimal, but heart felt. Images of drill and exploration work conducted by the company over the years hang on the walls throughout the office. Despite the company location, the office suite is quite small though functional. Three formal offices face the north side of the building overlooking the magnificent sails of Vancouver's Pan Pacific Hotel, while a small conference room and shared workspace area occupy the rest of the office. Due to the rapid increase in staff size over the year, employees have also claimed the break room as workspace.

It is clear that the company's main area of operation is out in the field, where the exploration work is being conducted. Much of Botha's time is split between the field and meeting potential and current investors. This frequent travel is reflected in his uncluttered office and in an ultra-light laptop that is ready to go at a moment's notice. Due to this schedule, arranging a meeting with Botha can be somewhat of a challenge— and if, his voice mail points out that he is "in the bush" most likely a response will be initiated by email, which he can answer when he reaches a location that has access to the Internet.

Communicating with Botha in person is more revealing than via email. His excitement about the business is immediately detected. Once asked about the business of mining, Botha warms up to the subject quickly. Getting straight to the point, Botha points out that the energy business is changing rapidly. One of the basic facts of the industry is that the use of oil for the production of energy is going to decline in the coming years. Not just because of the amount in the ground, but also because public opinion is turning against it. More people need more energy than ever before to support the use of modern technologies, but they want that energy to come with less carbon dioxide emissions. As Botha sees it, there are two solutions based on existing technologies: uranium for nuclear power plants and coal for clean coal or coal-to-liquid plants. Cash Minerals is involved in both of these opportunities, with the primary focus being the company's uranium properties located in the Yukon Territory, Labrador and British Columbia.

Cash Minerals' most developed uranium holdings are in the Wernecke Uranium District of the Yukon, about 120 kilometres northeast of Mayo. Uranium in the Wernecke district is part of a granite related hydrothermal ore deposit, also known as iron-oxide-copper-gold (IOCG) systems. In the summer of 2007, Botha visited the world's largest IOCG – the Olympic Dam mine – located in Southwest Australia, and was accompanied by Cash Minerals' newly appointed Vice President of Exploration, Dr. Geordie Mark. "There's a similar type of geology there, and we wanted to get an idea what we are dealing with," says Botha. "Very few people have actually visited the mine and been underground, so that in itself is an accomplishment."

While Canada's Athabasca Region gets much attention for its uranium because it has the highest-grade uranium deposits in the world, the Olympic Dam mine is impressive due its size. Thus, despite the fact that the average uranium grade is measured in kilogams per tonne in the Athabasca region compared to less than one kilogram per tonne at the Olympic Dam mine, it is still known as one of the world's largest copper-uranium deposits. The mine currently produces 4,500 tonnes of uranium oxide per year, and planned expansion efforts could take this as high as 14,000 tonnes per year. The mine also produces gold and silver, and is a major producer of copper, meaning that it is in an excellent position to respond to shifts in world markets and remain highly active. The proven and probable ore reserves of approximately seven billion tonnes are expected to keep the mine in operation for the next 200 plus years.

The Wernecke Uranium District has the potential to be the next major discovery in Canada given the geological similarities to Olympic Dam. Many geologists believe that the uranium deposits of the Yukon and Australia were formed during the Proterozoic age, 1.6 billion years ago, when the same geological processes were taking place. This theory would explain the geological similarities of the two regions. The two areas "are like brother and sister coming out of the womb at the same time," Botha said. "In our minds, the Wernecke district in the Yukon is the new frontier as far as a major uranium-copper discovery is concerned, because these IOCG systems are associated with size."

"We have one of the best teams in the business from a technical standpoint, due to their experience with IOCG models, and you do not get a team like this unless they are very confident in you and your projects," says Botha. "We need at least 100 million tonnes to make a go of it, so our goal is to find the elephant, and we have had some very promising results in this regard."

The team's confidence is based on two very real advantages that not all junior mining exploration companies have: the ability to raise cash and significant drill core results to justify subsequent expanded exploration programs. In 2005, Cash Minerals restructured to focus more attention on uranium exploration and spent $1.5 million that year on exploration in the Wernecke Uranium District. With one drill, the company produced 3,000 metres and came up with some impressive intersections; approximately 75 metres of 0.07 percent uranium oxide and 1.88 percent copper at the Igor property – similar to Olympic Dam grades, if not better. The following year, Cash Minerals returned to the field spending $9 million on two drills, drilling a total of 8,000 metres. The 2006 results were so encouraging that Cash Minerals initiated a staking program that saw the company more than quadruple their land position in the Yukon. Today, Cash Minerals holds over 1,100 square kilometres in the Wernecke Uranium District, which includes the recently acquired properties attained through this year's friendly takeover of Signet Minerals Inc. The 2007 drill exploration program is Cash Minerals' largest uranium exploration program to date. Approximately $17 million has been budgeted for exploration this year, which will employ six drills and a geological staff of over 35 people.

Another advantage of mining in the Wernecke district is that work has not been carried out here for over 20 years. Botha points out that in the 1970s, the last phase of strong interest in uranium in the region, the players were Chevron, Aquitaine and El Dorado Nuclear – companies that explore where real prospects reside and where the probability of success is high.

This year, Cash Minerals' drilling is focused on two properties in the Wernecke Uranium District: Igor and Lumina. Cash Minerals' Wernecke properties are under a joint-venture arrangement with Mega Uranium Ltd., a well-known Canadian resource company focused on the exploration and development of uranium properties. Currently, Cash Minerals and Mega Uranium each have a 50 percent interest in the Wernecke properties, however, Cash Minerals has the ability to earn up to a 75 percent interest in one or more of these properties.

Igor is an iron-oxide copper-gold-uranium target and was the first property among Cash Minerals' uranium holdings in the region to demonstrate potential. In 2005, results from Igor showed high levels of copper and uranium oxide. Comparisons of the mineralization, gravity and magnetics at Igor show a very similar profile to that found at Olympic Dam. Three drills are working at Igor this season, testing a significant copper-rich uranium intersection of 74.44 metres (of 0.07 percent uranium oxide and 1.88 percent copper), that was discovered during the 2005 exploration season. An underground gravimetric anomaly that was identified from extensive ground geophysical work conducted during the 2006 season will also be tested this year. If successful, drill results, gathered from the Igor property this year, will be used to define a resource and construct a geological model that determines the extent and continuity of the uranium–copper–gold mineralization..

The Lumina property, the other focus of the 2007 program, follows a different geological model from Igor. Lumina has vein-controlled mineralization that show uranium-rich veins less than 90 metres below the surface. Significant intersections achieved during the 2006 exploration season, included approximately 55 metres of 0.103 percent uranium oxide, which included 27 metres of 0.203 percent uranium oxide – economical grades. The main zone at Lumina consists of a 150-metre by 1400-metre long float train of highly radioactive boulders that lie atop an alpine glacier in a north-facing cirque. The area just above this float train is considered highly prospective and has been named the Jack Flash area. Representative samples of the radioactive material returned between 1.36 percent to 7.67 percent uranium oxide. This year the company drilled a series of follow-up holes to test the continuity and extent of uranium mineralization that was encountered last year. Two drills were allocated to Lumina to test the 2006 drill results and to follow up drilling in the south and the north. Additional work included field mapping, rock chip sampling and drill target definition, as well as high-resolution grids of soil geochemistry and scintillometer readings. Results from the 2007 exploration program conducted at Lumina shows the presence of significant polymetallic (uranium-molybdenum-gold) mineralization, thus enhancing the metal potential of the Lumina property and more importantly the Wernecke area.

The Odie property, which is located in the Wernecke District, was also drilled in 2007. Odie was discovered during the 2006 exploration program and is the largest property in the company's portfolio of uranium properties. Odie is located approximately 35 kilometres north of Igor, and, unlike Lumina and Igor, it is situated in a low-lying area that is hospitable to drilling for most of the year. Like Igor, Odie shows similar geophysical characteristics to the Olympic Dam mine. Drilling conducted during the 2007 season confirmed two anomalously copper-rich intervals. Anomalous levels of gold tend to occur within zones of elevated copper, and in IOCG systems, uranium tends to occur in association with these two metals. In the fall, one of the drills currently in use at Bear River camp will return to Odie to determine the extent of these occurrences.

One of the six drills secured by Cash Minerals was allocated to the Bear River camp also located in the Wernecke district. A 16-person team was established to conduct greenfields exploration, as well as to follow up on ground work conducted in 2006. Targets for drilling include the Bonnie and Vic properties, which are in the mature exploration stage and have geophysical profiles resembling those of Lumina and Igor, respectively. In addition, drilling will test the Angel, Bond, and Nad properties for the first time, all of which were discovered during the 2006 season.

At first thought, many think that the Yukon is a difficult place to work due to the weather conditions, but Botha points out that it is actually much easier to mine in cold weather than in the extreme heat encountered in places like South Africa or Australia. People are the most important resource by far, and it is far easier to keep them warm with the right clothing and equipment than it is to keep them from suffering the effects of intense heat. Due to the physical location of the properties (situated at different elevations), Cash Minerals is able to operate from mid-April through early November in the Yukon.

The three camps in the Wernecke District (Igor, Lumina and Bear River) employ a staff of approximately 100 people, and, although there is some rotation, approximately 85 people are in the camps at any one time. The onsite staff includes five geologists who hold graduate degrees, three of them at the doctoral level, overseeing a large team of students. Senior management is present onsite as well. Botha himself stays at the camps at least once a month and Peter Arendt, Vice President of Engineering, is in the field on a two-week rotational cycle to manage logistics.

Asked about future plans, Botha says, "If we achieve our objective this year of defining a resource – getting to where we can put together 'x' amount of pounds in the ground – that will be the beginning. We will then go to the market for more funds to continue exploration so that we may expand the resource number in the 2008 season. There is a lot of attention on the Yukon right now, and our goal is to find something of significance."

In addition to working in the Yukon, Cash Minerals entered a joint venture with Cornerstone Capital Resources in January 2007 to explore the Aillik uranium property located in the Central Mineral Belt, Labrador. This area is of intense focus for uranium exploration in Canada because of its proven resources. Cash Minerals is conducting airborne geophysical surveys and fieldwork this year and intends to drill next year. Aillik is highly promising as it is adjacent to more mature properties that show well-defined uranium resources. The Aillik property plays several strategic roles for the company, namely, extending the exploration season beyond what is feasible in the Yukon, and diversifying holdings both geographically and in terms of exploration maturity.

This year, Cash Minerals has been heavily focused on the potential with their uranium properties, but the company is continuing to keep itself diversified in energy projects. Cash Minerals' mature coal property is also in the Yukon. The Division Mountain coal deposit is 90 kilometres northwest of Whitehorse, and 280 kilometres by road from a year-round tidewater port in Skagway, Alaska. The company conducted extensive exploration on the project from 1972 to 1999 and had a scoping study completed by Norwest Corporation in 2005.

"Norwest's study indicates the potential for a 22–year open-pit mine life that could produce approximately 1.2 millon tonnes of Bituminous "B" coal on an annual basis for sale to the thermal market within the Pacific Rim," Botha explained. "An additional 175,000 tonnes could be available for a proposed mine mouth power plant located on or near the property. The study's authors estimate capital costs to develop the mine of $31.9 million, an internal rate of return of 59.6 percent and a net present value of $74.8 million."

Over its 20-plus years of working on the Division Mountain site, Cash Minerals has not only identified a valuable resource, it has also developed valuable relationships that will help bring the resource to market. Staff has worked closely with the Yukon government, the city of Whitehorse and the Champagne and Aishihik First Nations, whose property borders with the Division Mountain property. There is strong interest in developing joint-venture agreements at each of these levels of government, and this year Botha put forward a proposal to the territorial government to develop a 52 megawatt coal-fired power plant on site. The primary strengths of this proposal are that the main area of exploration in Division Mountain parallels the Yukon Energy Corporation electrical transmission grid, which would be used in the economic development of the area as new mines located in surrounding areas move closer to production.

A disciplined strategy, such as the one being used by Cash Minerals, requires a strong team to execute. Although Botha's 30 plus years of experience in the mining industry are clearly a crucial part of the overall strength at Cash Minerals, Botha chooses to be relatively modest about his own contributions, instead pointing to the achievements of his newly assembled geological team. Dr. Geordie Mark, Vice President of

Exploration, is an expert in the type of systems Cash Minerals is exploring. As Botha puts it, "Almost anyone who really knows IOCG systems is either old, dead or about to die; nobody is in their forties and fifties. Geordie is the exception. He is young – 35 years old – and has 12 years experience working with iron-oxide copper-gold systems in Australia, Canada and Europe."

Mark trained under Dr. Murray Hitzman, who co-authored the first paper on IOCG systems in 1992, and wrote one of the first papers on the geologic origins of Australia's Olympic Dam (IOCG) deposit in 1993. Mark has written and contributed to numerous papers in international journals and technical papers and reports on IOCG systems, and has also conducted training courses and workshops on IOCG and related systems. His 12 years of on-the-ground experience were spent on the exploration and interpretation of IOCG deposits in Australia, Europe, and Canada, and Mark has specialized in the genesis and interpretation of hydrothermal IOCG geological systems, such as the one Cash Minerals is working on at the Igor site.

The exploration team also includes Dr. Michael Carew, PhD and Dr. Damien Foster, PhD, who are both specialists in IOCG deposits; Raul Sanabria, MSc, who is a specialist in structurally complex mineralized fault systems; and Rick Zuran, who specializes in regional exploration and has more than 17 years experience working in the Yukon.

Peter Arendt, Vice President of Engineering, plays an integral role overseeing operations and logistics in the field. Arendt is a senior mining professional with 20 years of industry experience. He spent 15 years in mine operations and engineering roles for major mining companies in Australia and Canada. Arendt is a Professional Engineer with a Bachelor of Engineering (Mining) and a Graduate Diploma in Business.

Gregory Duras joined Cash Minerals as Chief Financial Officer in June 2007, bringing with him more than a decade of corporate and project finance experience in the resource sector. Prior to assuming this role, he held senior finance positions at various publicly listed resource companies. Duras is a Certified General Accountant and a Certified Professional Accountant, and holds a Bachelor of Administration from Lakehead University.

The board of directors at Cash Minerals is equally impressive. These board members include Chairman, Stan Bharti and directors William Clarke, Andy Rickaby and Peter Rowlandson.

Stan Bharti is a Professional Engineer with over 25 years of domestic and international experience in mine engineering, operations management and finance. His experience in public markets includes the acquisition and restructuring of companies in Europe, Australia, and North America. Over the last 10 years, Bharti has raised over $400 million in public markets. Bharti's leadership role with the successful Desert Sun Mining Corp. assisted the company in achieving its development and production goals at the Jacobina Gold Mine in Brazil. Mr. Bharti holds Masters Degrees in Engineering from Moscow, Russia and University of London, England.

William Clarke is an expert in international trade and investment, with extensive experience as an advocate for Canadian industry. From 2000 to 2004, he served as President and CEO of the Canadian Nuclear Association. More notably, Clarke served in Canada's foreign service for 34 years with assignments around the world, such as Canadian Ambassador to Brazil, Sweden and the Baltic Republics. He retired from the public service in 2000 as Canada's Chief Trade Commissioner, leading and directing Canadian government personnel worldwide to promote export growth and investment inflows. He is currently on the Board of other international companies and provides consulting and advocacy services for companies as a member of Trade Commissioner Consulting Services Inc.

Andy Rickaby is a professional engineer with over 40 years operations and management experience with companies such as Inco Ltd. and Dennison Mines Ltd., including uranium mine evaluation in Canada, the USA, and Kazakhstan.

Peter Rowlandson is a professional engineer with extensive experience in the mining industry. He has held a number of senior engineering and management postings throughout his 34-year career with large mining organizations, including Rio Algom, Inco, Pamour, Canamax, and Teck Cominco. Most recently, Rowlandson was General Manager for the Teck Cominco-Barrick Williams and David Bell Mine Hemlo joint venture. Rowlandson brings expertise in all aspects of mining from construction through to production management with particular success at improving safety, increasing mine production, and lowering operating and administrative costs.

In a sense, Cash's team echoes the Yukon's early mining days. When George Carmack filed a claim on August 16, 1896 that set off the Klondike Gold Rush, he had already been living in the Yukon for ten years, trading, fishing, trapping, and, oddly enough, working on a major coal deposit that he discovered near the village now known as Carmacks. Today, many historians think that Carmack was in the party that found and staked that gold because of his strong relationships with members of the Tagish Tribe, who already knew the area well. Like George Carmack 110 years ago, Cash Minerals has already been in the Yukon for quite some time. The company's work at identifying coal resources from 1972 through 1999 gave them a chance to gain the knowledge of the area and to build the relationships that put them ahead of the pack when it came to looking for uranium.

"There's strong potential for a world-class discovery in this area," says Botha, who likes to refer to a potential uranium find in the Yukon as an "elephant" because iron-oxide copper-gold systems, are often known for their enormous size. "Our 2007 program is the largest uranium exploration program in the Yukon. It puts us about two years ahead of everyone else there."

Experience and a practical plan spelled success in the Yukon gold rush, and are still the main factors of success in this new uranium rush 100 years later. With knowledge of the unique Proterozoic age geology shared by both the Wernecke district of the Yukon and the giant Olympic Dam mine in Australia, a world-class team that knows what it has to do and can do, and promising exploration results produced in three consecutive years, Cash Minerals is poised to discover the elephant.

Dejour Enterprises

obert L. Hodgkinson considers himself a successful commodities financier. He has a feel for economic cycles and the strategic importance of energy resources. He has shown time and again, over the span of his 30-year career in the energy sector, that timing is everything. Knowing when to act is as important as knowing where, he says, and that goes double for the uranium space. So, Hodgkinson is more than a little proud of the mining powerhouse he's helped create in just a few years by selling some of Dejour Enterprises' assets to Titan Uranium. It must also be especially gratifying to Hodgkinson since he started his foray into uranium with a modest $700,000 investment.

The Dejour offices on West Hastings in Vancouver are furnished in a utilitarian, no nonsense way. Gaining access to the company's inner sanctum follows the strict kind of security protocols normally employed in data rich oil and gas company offices in Calgary, and for good reason: While Dejour began as a uranium only company, it is now an active oil and gas explorer with a lot underway, and that effort is now augmented by serious ownership interests in uranium.

Before Hodgkinson picked it up, Dejour Enterprises was a moribund uranium exploration company originally founded by a mining industry giant named Duncan Ramsay Derry. Derry had sprung from illustrious stock. His mother was a daughter of William Ramsay, one of the founders of the Imperial Bank of Canada, which later became the Canadian Imperial Bank of Commerce. His father, Dr. Douglas E. Derry, was professor of anatomy at the medical school of the Egyptian University in Cairo. In 1923, acting at the behest of his friend, archaeologist Howard Carter, Dr. Derry undertook the autopsy on the cadaver of the Egyptian Pharaoh Tutankhamen.

Born in England, Duncan Derry was educated at Cambridge and later continued his studies in geology at the University of Toronto where he completed a PhD. Over the course of his career he became one of the most distinguished geologists in the country, receiving a number of national and international awards as well as an honorary degree from the University of Toronto. He was appointed an Officer of the Order of Canada, becoming the first geologist to receive that honour.

Just at the beginning of the second boom in uranium exploration in Athabasca in the 1960s, he and a Native prospector friend named John decided to create a company. They called it Dejour—De for Derry, Jo for John, and Ur for uranium. They raised money on the Toronto Stock Exchange to go prospecting in the Athabasca Basin, but history shows they were short on luck.

It's a different situation when Hodgkinson recalls everything that transpired in the short time he has been involved with Dejour. He gives his head a gentle shake as he ticks off the most notable accomplishments on his fingers. It's obvious even he is a little bit awed by the sum total of activity because a lot has happened since 2004. Through Dejour, Hodgkinson has become personally linked with Canada's historic third uranium exploration boom, just as he has already made history in another energy sector exploring for oil and gas. Through Dejour, he's had the pleasure of befriending one of the world's most revered uranium experts. And finally, through Dejour, he has had the chance to work with the hand-picked team of exceptional uranium explorers he believes could discover the next huge uranium resource in the Athabasca Basin.

As far as Hodgkinson was concerned, when he bought up control of Dejour the company represented an ideal vehicle to expand his energy commodity portfolio. As he moved to acquire the company in 2004, TSX mandarins were taking steps to have Dejour de-listed. But, relying on his experience anticipating trends, Hodgkinson saw a bargain waiting with Dejour and he gambled on the junior exploration company.

Hodgkinson is no stranger to taking a flyer on future outcomes although his risks are always well considered. For example, he was an early stage investor and original lease financier in Synenco Energy's Northern Lights Project which today has 1.5 billion barrels of oil in the reserve category in the Alberta Oil Sands. He was the founder and chairman of Optima Petroleum, a company which would drill 175 wells in Alberta and the Gulf of Mexico before merging to form Petroquest Energy, a NASDAQ-traded company with a market capitalization of $470 million USD. And, he was the founder of Australian Oil Fields, a petroleum company which would later merge to become $400 million Resolute Energy/Cardero Energy Inc.

"Dejour was nothing more than a shell with a really good pedigree," Hodgkinson says, but Dejour had 35 years of audited financial statements "and it was extremely clean." That attracted Hodgkinson. "When I decided to reactivate it and started putting the Dejour team together, I was really thinking about getting back into the oil patch," he says.

Hodgkinson attracted Doug Cannaday to Dejour in 2003 and appointed him President and COO in 2004. Cannaday has over three decades experience in oil and gas exploration. He has been a director, officer, or controlling principal for a number of exploration companies including Artesian Petroleum Corp., Amador Resources Ltd., Seahawk Oil & Gas Ltd., and Lava Cap Resources Ltd. In 1999, Cannaday founded Riomin Resources S.A. in Ecuador and subsequently served as a consultant for several Alberta oil and gas companies before joining Dejour.

Hodgkinson also attracted 30-year oil patch veteran Charles W.E. Dove as President, Dejour Energy (Alberta). Dove began his career with Amoco Canada Petroleum Co. in 1978. Before joining Dejour, he held positions with CDEC Oil and Gas Ltd., Diamond Shamrock Exploration Ltd., Quintana Exploration Ltd., and Rustum Petroleum. He was a co-founder and major shareholder in Innovative Energy Ltd., which was sold to Dennison Mines in 2001.

Matthew Wong became CFO of Dejour, Phil Bretzloff became vice-president and general counsel with the advice of Craig Sturrock as a director. Dr. R. Marc Bustin also became a director, filling out the oil and gas side of the energy spectrum at Dejour. Bustin brought 30 years experience in oil and gas exploration to Dejour too. He had worked with Mobil Oil Canada, Gulf Canada Resources, ELF-Aquitaine (France), CSIRO (France), and CNRS (Australia). Once a professor of petroleum and coal geology in the Department of Earth Sciences at the University of British Columbia, Bustin is credited with publication of 150 scientific articles on fossil fuels.

The oil and gas team at Dejour is impressive, but when Hodgkinson took over Dejour, he says "oil and gas stocks had taken off and the property prices were ridiculous. Also, at the time I hung up my shingle I noticed a lot of people from the United States coming up and dusting off their old uranium properties." Many were offered up to Dejour for acquisition, but Hodgkinson knew he had to be cautious. Sensing a staking rush in the offing, he decided to seek some advice and turned to one of the most revered experts he could find in Canada's uranium space.

Dr. Lloyd Clark has more than 50 years experience in uranium exploration and is recognized as one of the world's foremost geological engineers in his field. While he served as Exploration Manager and Chief Geologist at the Saskatchewan Mining Development Corporation (now Cameco) he established the Exploration Branch and for nine years was in charge of a 65 member geological staff. Among other achievements, Clark's team at SMDC made the McArthur River discovery, which is currently the world's largest, most profitable, and highest grade uranium mine.

"I retained Lloyd and, through Lloyd, three or four other people to review all the deals I was being offered. There were about 30 of them." In the end Dr. Clark nixed each of the offers, "saying 'if you're really serious about this I'll bring in all my data. I think you should go right back to the Athabasca Basin. That is where I think we should make our stake in the uranium business.' So he brought all his data in and we compared what he had and what we knew was leased in the Basin." After choosing some likely targets, "we appropriated a million dollars and put helicopters in the air in October and November of 2004. By the end of January we had about 700,000 acres and by March 2005 we had about a million acres in the region for $1 million CDN. It was at a time when the spot price for uranium was $14 to $16 per pound. That is how we got started."

In making his moves into the Athabasca Basin, Hodgkinson had another beneficial ally in Archibald Nesbitt. With over 25 years experience in the development and financing of junior resource and venture companies, Nesbitt is a Dejour director with some personal knowledge of the territory Dejour was staking. His career began in 1966 when, with his late father John C. Nesbitt, he focused on exploration for uranium near Uranium City, Saskatchewan. John C. Nesbitt had discovered the Nesbitt, Labine, and Gunnar uranium mines there two decades earlier.

"Archie's father was a bush pilot and his discoveries spurred the first of what became three major uranium exploration booms in northern Saskatchewan," says Hodgkinson, outlining Dejour's corporate history. That story begins with tales of uranium mining in Saskatchewan that extend back to the mid-1930's when pitchblende mineralization was discovered around the settlement of Goldfields on the north shore of Lake Athabasca, and when Gilbert Labine discovered massive pitchblende deposits near Great Bear Lake.

In 1943 the federal government in Canada imposed a ban on the staking and mining of radioactive materials by the private sector, leaving that entirely as the duty of a Crown corporation called Eldorado Nuclear. By 1944, Eldorado was hard at the exploration and trying to identify and develop new sources of uranium. Compared to modern exploration tools, however, the methods for discovery Eldorado's prospectors used were downright simplistic. They roamed over the Athabasca Basin waving handheld Geiger counters. Even so, they still managed to find more than 1000 pitchblende showings during the 1945 field exploration season.

The following year, Philip St. Louis discovered mineralization near a significant geological fault zone in the area. The recognition of the St. Louis Fault pre-staged the development of Eldorado's Ace, Fay, and Verna mines. In 1948 the government lifted the ban on private sector staking and the search for radioactive minerals expanded. Incentives were offered. The Saskatchewan government set up prospectors' training schools to encourage uranium prospecting and practically overnight one of the largest staking rushes in Canadian history began in the northern reaches of that province.

Exploration in the Athabasca Basin in the 1950s was based on the vein model. That model led to the discovery of the Uranium City mining camp where about 65 million pounds of uranium was mined from vein deposits. The town of Uranium City was established by 1952 and at its peak the northern mining community claimed 3500 residents. In the next the three decades, however, more than 200 companies staked claims as well. A total of sixteen mines were brought into production with the Ace, Fay, and Verna mines operated by Eldorado consistently being the most prolific and producing over 40 million pounds of uranium alone.

"In the 1950s, mining was really an Ontario/Quebec jurisdiction," says Hodgkinson. "One of the most pre-eminent mining engineers in the country was Duncan Derry. Bay Street in those days was mostly a mining street and he knew everybody." But the odds of discovering a uranium deposit, developing it and then exploiting it to mine stage are daunting. Estimates by a leading mineral exploration company once put the chance of successful discovery and development at roughly 1:10,000. Success is akin to finding rich mineralization the size of a football field in an area 80,000 square kilometres in size. Not surprisingly, the exploration activity by the Dejour team of the 1960s, like that of all but a handful of other explorers, came up dry.

By 1968, exploitation of the deposit at Gunnar Mines ended. Eldorado was the only uranium production company operating in Saskatchewan until the cap came off the genie bottle once again: Another major uranium discovery in the Athabasca Basin was made at Rabbit Lake that year. The Rabbit Lake discovery was based on a roll front model rather than a vein model. Its discovery led to a uranium rush that, in turn, resulted in a high grade uranium deposit being discovered near Cluff Lake, Saskatchewan in 1969.

To encourage even more exploration in the province, in 1974 the Saskatchewan government created the Saskatchewan Mining Development Corporation and the developmental strategy worked. Within a year the richest open-pit uranium deposit in the world was discovered near Key Lake, about 250 kilometres north of La Ronge. The Key Lake discovery, and other discoveries that followed, were the base for the unconformity uranium deposit model being used by explorers today. Over the course of 25 years, the Rabbit Lake Mine produced 120 million pounds U3O8. Comparatively, over its 15-year life, the Key Lake deposit produced more than 190 million pounds U3O8.

Using the unconformity model since 1968, geologists have discovered 18 deposits totalling over 1.4 billion pounds of uranium in the region. In the early 1980's, the MacLean, Midwest, and Sue Deposits were discovered, as was the enormous, high-grade Cigar Lake Deposit. Cigar Lake, at 300 metres depth, hosts reserves of 551,000 tonnes grading 19% U3O8, or 232 million pounds U3O8.

By the mid-1970s it was apparent that the high grade surface targets were gone and geologists began to hunt for lower-grade deposits, 0.3 to 1% U3O8, which could be exploited by standard open-pit or near-surface extraction techniques. That meant targeting the edges of the Athabasca Basin and much of the Basin rim was therefore explored.

By the early 1980s, however, explorers determined that as well as the shallow basin edges, uranium could be found in deeper sections, areas where the sandstone exceeds 400 metres thickness. Technology ramped up quickly. By the late 1980s, airborne surveys were employed to detect conductive structures at ten-times the depth below the Athabasca Sandstone than was possible just five years earlier. In 1988, the highest-grade uranium deposit in the world was found beneath 600 metres of sandstone at McArthur River. The McArthur River Mine entered production in 1989 and currently produces 18 million pounds annually with a mine life expectancy of more than 20 years. It is estimated to hold 235,164 tonnes of uranium with an average ore grade of 19.60%.

These later discoveries have no surface expression so they are known as blind deposits. Exploration for blind deposits requires geologists to combine their knowledge of geological models derived from earlier discoveries. It also involves the application of deep sensing geophysical techniques and skilled interpretation of geochemical signs and signals associated with these unconformity-type deposits. The deeper, blind deposit discoveries however, prove there may still be opportunities in the Basin. With tools like electromagnetic geophysics, explorers can re-examine areas that were passed over in the past. It was use of these EM geophysical methods which identified graphite in the basement fault structure at McArthur River. That insight helped guide a drilling program that located the main ore zone. It is precisely that type of high-tech prospecting that companies now hope will help them find the next McArthur River-like uranium elephant.

"Dr. Lloyd Clark is probably one of five historical geologists who really developed the theories we use today about why there is so much uranium in the Athabasca Basin, how it got there, and where you should find it. Lloyd has a data set that probably only a half dozen others have. When he brought out the old maps from 1970 we saw Dejour everywhere on the maps of the Basin, but as it turned out they had never found a thing," says Hodgkinson.

However, after acquiring Dejour, Hodgkinson relied on Dr. Clark to help him build an exploration team. Dr. Clark has spent ten years as a professor of geology and geochemistry at McGill University in Toronto, and three years as a pre- and post-doctoral research fellow at The Carnegie Institute in Washington, D.C. Geophysical Laboratory. "We hand picked a team. Most of them had been students of his and all had worked for major companies at one time or another. We set about doing what had never been done on these properties, which was re-mapping all the data. Then we compared that with the data on other properties because they are geologically intertwined and the trends we were studying don't have any geographic boundaries. Of course for many of these properties the data we studied was from the first boom and some of it was from the second boom. But the technology got better and better. Throughout the 1990s and through to the current decade the technology looks better and looks deeper, and it can give you a better idea of what's going on. So, we did about $7 million of advanced geophysics and came up with a whole data set that complemented or condemned the older data and that really started driving us in other areas.

"We focused more on the western side of the Basin because that is where Lloyd, Cameco, and a few others feel is the next big deposits are going to be. That is the area of the Basin that is the least explored and it is a little deeper. Historically there was a very significant discovery south of Cluff Lake by UEX and Cameco has made a very important discovery too. It was around those discoveries that we started to focus our land acquisitions.

"At that time those deposits or discoveries were not as advanced as they are now. That is how we got going. It's a very interesting story because it has all the synergies of history and people. We were very fortunate to be there early in the third boom and to be able to raise the kind of money we did from the market. Because of that we were completely focused from 2004 through 2005 and the program was really accelerating."

In 2006, with uranium prices rising to $70–$75 per pound, the Dejour Board of Directors decided it was time to monetize the company's holdings. "That's when we began to say to ourselves: 'Boy there are a lot of junior uranium companies coming along right now. It would be nice to hedge our bets.' We started looking for a potential partner." Recognizing the protracted nature of uranium exploration, Dejour's management began to seek an optimal partner to leverage the company's holdings and mitigate the rising cost of exploration.

Hodgkinson had never completely taken his eye off the oil and gas sector either. While Dejour was actively acquiring land in the Athabasca Basin, the price of natural gas peaked and was going south. The price of oil had come off $70 per barrel and was trading in the low $60s. Oil and gas was looking a bit vulnerable to the energy commodities financier. Hodgkinson noted the amount of exploration activity by major oil companies in the Piceance and Uinta Basins straddling the Colorado-Utah border, and he decided to act in 2006. Dejour acquired 288,000 gross (60,000 net) acres in the Piceance and Uinta, the most prolific oil and gas prospect in the Western Hemisphere.

It proved to be another case of the "right place and right time" for Hodgkinson. The acquisition, which cost Dejour about $25 million (USD), gave the company a 25% retained interest and 12.5% retained interest. Since that acquisition, major finds in adjacent properties by Delta Energy and Exxon Mobil have increased the value of Dejour's holdings substantially. Recent estimates suggest undiscovered resource potential for these holdings of 4 trillion cubic feet natural gas and over 2 billion barrels of oil.

"The Piceance is very comparable for oil and gas to what the Athabasca Basin is for the uranium business, albeit the Piceance Basin is much less risky than searching for uranium. That's when we started contacting other juniors," Hodgkinson says. "There are certain companies that had synergies in other areas. We thought they could evolve into significant uranium camps of their own in the Athabasca Basin. One of those companies was Titan Uranium."

After initiating some discussions, Hodgkinson says Titan's management was impressed by the Dejour team and strategy. "They said, 'We like what you are doing and think you should buy us.' We said "that would put us in a greater conflict strategically. Maybe what we should do is take our assets and make you bigger.' They had just hired Phil Olson, a very respected mining executive, and subsequently they brought on Brian Riley who was the second-in-command at the AREVA Group subsidiary, AREVA Resources Canada Inc.," Hodgkinson recalls. For Hodgkinson, the marriage between the companies was a 'no brainer'.

"With their geophysical team and our exploration team merged in the same place Titan has become one of the most competent exploration teams in the Basin. It is a team that really understands the Athabasca Basin and can implement an exploration strategy there." He feels the deal has made good sense for all concerned. "We had avenues to raise capital that they didn't. We had offices in Saskatoon as well, so now they have one office with all the intellectual property under one roof, managed by a very cohesive team. They've got well over $20 million in the bank and they've got another $14 million in joint venture funds.

"From a Dejour point-of-view, we had no exposure in the Thelon basin and no expertise to get there, so that came as a freebie with the deal." He says the knowledge that Riley can add to the mix for Titan is considerable as well. AREVA Resources Canada Inc., owns and operates the Cluff Lake mine. It is the operator and majority owner of the McClean Lake and Midwest uranium projects and part owner of the Cigar Lake, McArthur River, and Key Lake uranium projects.

In December 2006, Dejour closed its agreement with Titan Uranium Inc. to trade $7 million in exploration/property expenditures for $46 million in stock and warrants in Titan. Dejour also retained a 10% carried interest and a 1% NSR in its original projects.

The agreement immediately made Titan Uranium one of the largest uranium explorers in the Athabasca Basin. Titan emerged from the transaction with 1,440,000 acres of prime uranium exploration lands at the world's 'Number One' address, a geological team utilizing 230 years of professional experience on uranium discovery in the Athabasca and Thelon Basins of northern Canada, and the working capital to implement. "Very interestingly in the Athabasca Basin, which historically is the number one producing basin in the world, there has not been another major discovery for a long time. There are some leads and the extensions of older discoveries but you haven't got a new major discovery at this point in time. I think with the amount of money being spent on exploration there are the opportunities for that out there. We'll see some measure of success and we should start seeing it this year because you've got so many companies that have gone past the early stages of exploration and are now actively drilling like Titan."

The Basin's prolific record—eighteen uranium deposits discovered since 1968, grading from 0.3% to over 14%—certainly suggests potential for new finds. However, it will be geological know-how and access to a large supply of exploration cash that separates the companies making real gains from those simply riding the coattails of the Basin's reputation.

"I believe clearly that we are in the biggest energy commodity boom of our lives and it's got everything to do with the quality of life. Everybody wants to be where North America has been. Don't forget oil was found in North America originally (the Thorla-McKee discovery in Ohio in 1814). No matter what you say about energy, the more you have the more you use and the richer you end up being. The rest of the world wants it so the price is going to go up.

"The reason uranium took off so quickly is because the supply isn't there anymore. The supply for years and years was government residuals of the Cold War from the armament stockpiles and that is basically gone. And all of a sudden we are into a situation of global warming. As much as I think CO_2 is not the most realistic reason for what is happening to our world today, I definitely agree that nuclear power is a significant future energy source for us. It makes sense. Our technology has gotten to the point where we can depend on it."

Therefore, Hodgkinson does not believe the uranium market will wither in the foreseeable future. "The fact is there are going to be significant uranium discoveries and when there are discoveries, the price of those shares are going to do extremely well in the marketplace. Dejour owns about a third of Titan and we have about a 10% position in all those Titan properties in the Athabasca and Thelon Basins and that is a very good place to be. We continue to be very active with the Titan people, monitoring where they are drilling, how they're doing and just where they're going. I talk to Phil Olson continually about what is going to make Titan better, because for Dejour, Titan is a major investment."

"Here's my personal opinion about the future. Between now and 2015 you're going to see the price of oil go sky high. You're going to see natural gas at more than double where it is today and you are going to see uranium a vibrant and consistent commodity. I believe there will be some big discoveries but these discoveries are going to take a lot of money to put into production so the whole nature of the ownership of these uranium discoveries is going to change. I think you're going to see groups that are totally funded by utilities around the world owning these deposits in the end. That's going to be very, very good for teams that are sophisticated like Titan's team. I mean teams that can actually cultivate and have the kind of credibility necessary to get these kinds of deals done. I think Dejour is in an excellent spot here. From an investment point of view I think the next five to seven years is going to be a wonderful place to be. It's not a one way street mind you, but the direction is very, very clear."

To succeed as a junior uranium exploration company these days, Hodgkinson also believes companies have to be robust enough to act when the time is right. "You have to be able to weather the peaks. You need to have value-added on the way through but in the end it's all about discovery. You've got to know how to make these discoveries and you've got to be fortunate enough to actually make them. That takes a very disciplined approach to the business of exploration and the skill sets like those that Titan has today."

Hodgkinson reminds us that at the moment it appears that by 2015, Cigar Lake and McArthur River will be the only known deposits still in production for a total of 36 million pounds U3O8 per year. Eventual exhaustion of the presently known reserves for these two deposits necessarily requires that Athabasca production begins to decline sometime after 2020.

How rapidly Athabasca production declines depends on future development of either currently uneconomic resources, or new discoveries. And Titan, he believes, has the potential land position and the knowledge base to play an historic part in finding new resources for the world's future.

ESO Uranium Corporation

As Tony Harvey unfolds a four-foot by three-foot map identifying claims in the Athabasca Basin, his vice-president exploration, Ben Ainsworth, produces a sheared sample of a drill core. The heavy core has black flecks traced with mottled splotches of yellow. The rock contains uranium in the raw.

The map is checkered with large blocks of color, a good portion of which denote ESO Uranium Corp's nearly one million acres of property in the Athabasca Basin. The uranium-mineralized core sample Ainsworth offers for examination comes from the company's Cluff Lake project just a few kilometres from the once world famous mine site in northern Saskatchewan. Harvey places it on the map and points carefully to a spot on the colorful chart with an index finger. "Right there," he says smiling. "That's where we drilled it."

ESO stands for Energy, Solutions, Opportunities and the company ranks as the fifth largest landowner in the Athabasca Basin. Harvey, the Chairman of the company, has good reason to grin a little.

The Athabasca Basin is the world's most important uranium producing district and supplies almost 30% of the world's primary uranium. Uranium is used mainly as a fuel for power generation, an important alternative to carbon based fossil fuels such as coal.

Current production in the Basin is mainly from unconformity-type deposits, lying at the bottom of a sandstone-filled sedimentary basin that overlies the rocks of the Canadian Shield. The first mines were found towards the edge of the basin where the sandstone is thinnest. Since 1968, eighteen deposits totaling almost 1.5 billion pounds of uranium have been discovered in the area, including the rich McArthur River Mine with total reserves of 367 million pounds at an average grade of 20.55% U_3O_8. The earlier discoveries, including Rabbit Lake and Key Lake, were near surface and found through prospecting and ground geophysics. Later discoveries such as McArthur River and Cigar Lake were located using deeper penetrating geophysical surveys as well as a range of methods including boulder sampling, geochemistry, airborne and ground geophysics, and drilling. The overall dimensions of high grade deposits are very small. Half of the reserves at McArthur River are contained in a zone measuring only 70 metres long by 30 metres wide by 70 metres thick. Fortunately there is often a large alteration zone associated with the important mineralization. This means there is excellent potential to find more of these small, high-grade, blind deposits with the enhanced exploration database and advanced geophysical systems that are available today.

Ainsworth's uranium-rich core sample comes from the company's most advanced Cluff Lake exploration target in the Gorilla Lake area where ESO drilled seven holes during the winter 2006 season. Four groups of mineralized boulders were found in that area but the sources for them remain undiscovered. Harvey and Ainsworth say they think the sources are likely to be on the ESO Cluff project property and the drill holes may prove them right. ESO had significant shows of uranium in three of the seven holes it drilled. "We have already intercepted uranium mineralization on the Gorilla Zone. We drilled seven holes of which two intersected U3O8 at more than the world average grade of 0.2%," says Ainsworth. The holes intersected uranium mineralization and identified a structural zone more than 700 metres long. Hole CLU-01 came in at 0.48% U3O8 over 1.5 metres. CLU-07 encountered 0.17% U3O8 over 7 metres, including two higher-grade intervals: 1 meter of 0.82% U3O8 , and 2 metres of 0.20% U3O8. After a detailed review, the structure of the Gorilla Lake zone has been interpreted by ESO's technical team as being near vertical. It is believed that past drilling, which was all vertical, could have missed larger uraniferous zones. By drilling angled holes, ESO's objective is to obtain wider, more representative uranium-bearing intercepts. A further 100 hole drilling program is taking place over the summer and winter 2007/2008 seasons.

In the Western Athabasca Basin Area that adjoins the old Cluff Lake mine site, ESO controls 169,000 acres which includes approximately 144,000-acre under a 50/50 J.V. agreement. The former Cluff Lake mine, which produced more than 60 million pounds U3O8 during its lifetime, is surrounded on two sides by ESO property. A portion of the Cluff Lake Project is also strategically situated on strike and trend with the AREVA/ UEX Shea Creek discovery zone which is located 9 kilometres to the southeast. ESO has a 50% joint venture interest with Hathor Exploration Ltd. and has also completed the work commitment required to move to a joint venture agreement with International KRL Resources, and Logan Resources for continuing exploration on a 50-50% shared expenditure basis. These joint venture lands only amount to about 7% of ESO'S interest in 985,000 acres in the Athabasca Basin, with the remainder being a 100% interest to ESO.

The Cluff Lake Project property is in what is known as the Carswell Dome area. The Carswell Dome is interpreted as being an ancient meteorite impact site that penetrated through the Athabasca sediments into the underlying Pre-Cambrian basement rocks. Elsewhere in Saskatchewan, several high-grade uranium deposits have been found and mined at the unconformity at the base of the Athabasca Basin. It appears that the Carswell impact structure caused the re-positioning of uranium mineralization from this environment to form the high-grade deposits near the present day surface when a resurgent mass of basement rocks irrupted to the surface during the rebound that followed the violent impact. The Cluff Lake uranium deposits were brought to the present day surface location by this cataclysmic event.

ESO has another prime prospect property in which it has 100% control, the 130,624 acre Hook Project. Cameco, the previous operator at Hook Lake, had noted that, based on the occurrence of nine geophysical conductors with a strike length of over 78 kilometres, the potential to have a significant unconformity hosted uranium deposit somewhere on the Hook Lake claim area was high. Tony Harvey and Ben Ainsworth agree about that too.

In late August 2007, ESO announced the planned start of a ten-hole diamond drill program on the Hook Lake uranium claims as well. The Hook Lake Project adjoins mineral tenure of Cameco Corp. to the south, recently optioned to Purepoint Uranium Group Inc. To the west it adjoins claims held by Fission Energy Ltd., and to the east claims held by Titan Uranium Incorporated. All three areas are under active exploration at this time.

The ten-hole drilling program will test targets on a possible extension of the north-northeasterly trending Dirksen Structure, a uranium-bearing fracture zone which may extend onto the Hook claims. A second target is on a sub-parallel structure lying to the west of the Dirksen. Earlier work indicated the presence of geochemical and mineral signatures in drill holes near and around these targets. These kinds of signatures are often associated with the alteration plumes that can exist above high-grade uranium deposits. Only time will tell if the assumptions bear fruit for ESO.

The Mandin Project is another 61,505 acres 100% owned by ESO. Situated further east of the Hook Lake claims, the Mandin will be reviewed for further target development, based on the interpretation of extensive deep penetrating geophysical surveys carried out in 2006 and 2007.

ESO also has extensive holdings on the eastern edge of the Athabasca Basin consisting of a 100% interest in 54 claims covering 621,934 acres in another three projects. The Peterson and Cree projects are primarily to the northwest of the Millenium Deposit, the McArthur River and Cigar Lake mines, on the far eastern boundary of the Basin.

"In 2007, ESO engaged MPH Consulting, an independent firm with considerable experience to run our eastern Athabasca holding of approximately 625,000 acres," says Harvey. "We are doing an evaluation of these properties and we may bring in joint venture partners on properties that we think we may not be able to focus on, but that may be of interest for someone else to develop while we work on our most important properties. When you've got a million acres you can't do it all yourself. It is too expensive."

Tony Harvey, with his 40-year career, has been involved in the design and construction of fourteen mines world-wide with capital costs ranging up to or exceeding $400 million. Formerly with Wright Engineers Ltd – Fluor Daniels in senior management positions, he is also president and founder of ARH Management Ltd., a management and consulting company to the resource industry. He was a founder, director, and senior executive of Azco Mining Inc., a resource company formed in 1988 and trading on the Toronto Stock Exchange and the American Stock Exchange. He was also a director of Cobre del Mayo, a Mexican mining company formed in partnership with Phelps Dodge Corp.

Harvey started ESO by acquiring a TSX capital pool company with which he moved into the junior resource sector. In 2003, the company acquired 100% interest in the 26,885 acre Mikwam Gold Project. The Mikwam Project is located along the western extension of the Casa Berardi Deformation Zone, close to the Quebec border in Noseworthy Township, and west of Aurizon's Casa Berardi Mine. The property has an extensive history of exploration and cumulative drilling that is highlighted by more than 547 drill holes with associated exploration costs estimated between $11 million and $15 million. ESO has plans to resume drilling at the Mikwam Gold Project in the fall and winter of 2007, but by far the company's sharpest focus is currently on its extensive holdings in the Athabasca Basin.

"ESO is well funded and has started spending some serious dollars to execute the ideas developed from earlier work and new exploration surveys using new technology. That expenditure indicates that ESO must be a serious player and has generated a certain amount of investor interest in Europe. The 100-hole program currently underway is related to targets in the Cluff Lake area and further drilling is planned for targets located on ESO's other claims in the Hook Lake and Cree Lake areas."

ESO has completed both AeroTem II and Megatem geophysical surveys on most of the property it controls in Saskatchewan. "Rob Beckett, our Chief Geologist and Manager for Saskatchewan projects, and Ben have done a remarkable job researching all of the historical information pertaining to the Athabasca Basin," says Harvey. "They've gone back to the old files in the government of Saskatchewan, you don't want to spend money drilling until you've done your research, research is cheaper than drilling." And the research points to huge potential for ESO in the Athabasca Basin, with the most promising prospect at the moment being the company's crown jewel property on the western side of the Athabasca Basin—the Cluff Lake Project.

Amongst the historic drill holes on the north side of the property found by research, Mokta reported that hole CAR 425, cut 0.85% U3O8 over 2.3 metres. "That's pretty high grade," says Ainsworth, "when the world-wide average grade for uranium mine production is 0.2% U3O8." The Cluff Lake Project is north of the decommissioned Cluff Lake Mine and the more recent AREVA/UEX Shea Creek discovery.

ESO now has an exploration database for the Cluff Lake Project derived from over 40 years of previous exploration including over 600 drill holes, Geotem airborne surveys, and detailed ground geophysical and geochemical surveying and prospecting. This property covers the northern two-thirds of the Carswell Dome structure, host to the Cluff Lake uranium deposits.

"Our 2006 drill assays were reported by international speculator, Doug Casey, to be among the best in the Athabasca Basin that year. Furthermore, we are intersecting uranium at shallow depths, from surface to 160 metres, versus the average target depths of 300-900 metres." says Ainsworth. " In the Gorilla Lake area we drilled six holes of which two were into U3O8 at more than four times the world average grade and fairly shallow. We have started drilling a 100-hole program on the Bridle Lake area, which is situated to the south of the former Cluff Mine. We were drawn there by historical work done back in the 1970s where they identified uranium with a grade of 0.12% or 2.4lbs per tonne which makes a very interesting target for us at current uranium prices."

In July 2007, ESO announced it had completed the first group of drill holes in its 100-hole program near the former Cluff Lake Mine that would continue into winter 2007. Two of the first three holes in the Bridle Lake Zone each had intervals of alteration and anomalous radioactivity, ESO announced. A further group of seven drill holes was targeted at electromagnetic conductors approximately 1000 metres east of the Bridle Lake Zone in an area with similar magnetic signatures to that seen over the Cluff Lake Mine lease. All seven holes intersected conductive graphitic meta-sediments. The next ten holes targeted areas of radon anomalies, some of which are close to similar conductors.

Radon, which occurs as a highly mobile gas, is one of several natural decay products of uranium. Its distribution in soils and water has been used, together with other exploration methods, to focus on areas with higher potential for uranium mineralization. The largest of the radon anomalies lies up-ice of the radioactive boulders that reported assays up to 0.85% U3O8. The radon anomalies are indicating anomalous uranium sources either in bedrock or in the overburden and soils material and may be indicative of the source material for the boulders.

ESO has now completed the work commitment requirements that allowed the company to form a joint venture with Hathor Exploration Limited. To that end, in August 2007, the two companies signed a full joint-venture agreement with a 50/50 interest in the Hathor's Cluff Lake properties. This allows the continued exploration of the properties that were originally acquired by Hathor in 2004, early in the development of the current surge of exploration for uranium

ESO had made exploration expenditures of over $5.25 million on its uranium properties in the Athabasca Basin prior to the 2007 season and has budgeted a further $6.0 million for the continuing programs in 2007/2008. These will include the intensive drilling of near surface targets at Cluff Lake indicated by the first phase exploration results and follow-up ground and airborne surveys that have been started on the Hook claims. ESO's excitement about the Cluff Lake Project gets a further boost from the fact it has conductors within the southwest area of the property, striking northwest and southeast towards the AREVA/UEX Shea Creek discovery just 15 kilometres southwest.

The uranium sector has unlimited upside potential, in Harvey's opinion, "lets face it, the world is running out of cheap oil and since no other commercially viable energy alternative is in place, uranium can immediately fill that gap. We can not continue to use our oil resources the way we have in the past. What are the reasonable alternatives? Wind power is expensive. Water power may be a little more efficient and cost effective, but that doesn't leave you with many options.

"Electric power generation has to go nuclear as demand exceeds supply. When China started to emerge economically it was obvious to them that it was is going to need a huge amount of power. Reportedly there are plans for 40 new reactors in China. Similarly India has plans for 24 new nuclear reactors in the next ten to fifteen years. Russia originally said it was going to build nine new reactors and it has now moved that estimate up to approximately 30 new reactors. Even the U.S. last year took its estimates for nuclear plants up from 9 to 30."

With the high grade Cigar Lake Mine now delayed there is a further squeeze on uranium supply. More mines have to be found and put into production to meet the growing global demand for uranium. That places ESO, with its large land holdings in the Athabasca Basin, in an enviable position for the future.

ESO'S management team includes Ben Ainsworth as Vice President of Exploration and Director. His first uranium exploration experience goes back to 1959 in Devon and Cornwall, U.K. with the UK Geological Survey. Ainsworth is a geologist who has been involved in mineral exploration and mining for over 40 years. After graduating from Oxford University, Ainsworth worked for Placer Development for 20 years enjoying two more uranium exploration booms and holding positions of Senior Geologist, Chief Geochemist, Exploration Manager–Eastern Canada, Exploration Manager–Chile, and President–Placer Chile, South America.

Robert Beckett, as Chief Geologist and Exploration Manager, has more than four decades experience in exploration, development, and operations. He was District Geologist for Esso Minerals, responsible for the Yukon, NWT, Northern B.C., and Saskatchewan. He also worked as the District Geologist for Saskatchewan Mining Development Corporation overseeing the northern half of the Athabasca Basin for a time. Beckett has also been exploration manager at Midwest Lake Mine and chief geologist of the Port Radium Mine. Before joining ESO, Beckett discovered gold mineralized outcrops and managed follow-up exploration that outlined over a million ounces of gold now in production as Goldcorp's San Martin mine in Honduras.

Jonathan George joined ESO's team as CEO, President and Director in 2003. George's extensive corporate experience included being active in numerous early state exploration projects through Canada, the U.S., Mexico, South America, and Northwest China.

Kurt Bordian joined the Company as CFO in November 2006 and brings more than 15 years of experience to ESO in the financial and accounting fields, including experience with numerous publicly traded companies during the past decade.

Independent advisors and Director's include, James Yates and Edward Marlow. Yates, as president of Hycroft Resources from 1982 to 1988, was instrumental in finding the Crowfoot Mine, a gold heap leach operation situated in Nevada, with annual production of 100,000 ounces of gold. Yates has been involved in the corporate management and financing of a number of projects in North America including American Bullion Minerals, Zappa Resources, and Jersey Goldfields. During his career he has raised in excess of $20 million for mineral exploration development.

Marlow, Managing Director, Principal Investments at HSBC Investment Bank, London, was previously with Anglo Dutch Merchant Bank, Insinger De Beaufort, and in asset management for UBS and Citigroup. He holds a MBA degree and a post graduate degree in Law. A former British Army Officer with considerable operational experience, a graduate of the U.S. Army Command and General Staff College, Marlow's military training is an obvious asset to a company with so much mineral lands to explore and a strategy to hunt for uranium.

The ESO team covers the bases of good corporate governance, the ability to build adequate financing and the technical depth to carry out the exploration. "We have developed new ideas based on better understanding of the uranium deposits and nowadays we have more sophisticated equipment. We can go back to where people had sniffs of minerals and see whether or not they missed something. You never take anything for granted. You have to recheck. How many people have walked away from a property and said it isn't a mine and have been proven wrong because they ran out of new ideas?" Harvey chuckles.

Considering the ESO Cluff Lake Project property for example, the team is debating the most viable production methods now. "We have very shallow targets and in fact, we are wondering whether or not in certain areas we can do in-situ leaching." Harvey thinks the team might be able to apply formation-fracturing experience that has been gained in the oil and gas industry to assist the exploitation of an in-situ leach deposit. "You always want to be leading edge and to think ahead. There are too many old miners who say 'we've been doing it this way for the last 50 years so why change?' Well you have to change," he says philosophically.

"There are many challenges to overcome in achieving something like that," he adds. "One is manpower. The other is equipment. In the Industry we have this great gap between experienced people who are still energetic and the new crop of geologists out there. It's hard to find people with experience." That is why Harvey feels so fortunate to have people like Ainsworth and Beckett on his team.

Ainsworth believes that ESO would be further ahead in its exploration cycle if not for the heavy demand now placed on drilling equipment and the people who operate the rigs. "The lack of drills is a problem. 'Two weeks' are probably the worst two words we've heard. It's always 'we'll be ready in two weeks' and then the experienced drillers and helpers are limited in number so, once a drill has been found, the manning of it with skilled crews is the next problem; also it's been hard to find competent field geologists with uranium experience. We are very much into fast mentoring. ESO has had great support from its drill contractor, Cyr Drilling, on the Cluff and Hook projects this year and although drilling costs are higher now this is cost that is spread across the industry in all service sectors by the demand created from the commodities boom."

"The Saskatchewan government staffing limitations initially had a dampening effect on ESO's exploration progress but this has since improved a bit. We tend to want to run faster than we are allowed. For a time, the Saskatchewan government had just one fellow approving permits and that created time hurdles", Harvey explains. "But even so, Saskatchewan is one of the friendliest places for uranium mining companies I know. You are never without some challenges no matter where you go, but I really feel very comfortable in Saskatchewan." Both the political and the geological environments in Saskatchewan are among the most positive and safe in the uranium business.

ESO has been working hard to develop strong working relationships with the First Nations people living in the company's project areas. "This is a very important part of what we have to do. We keep an open dialogue with other operators too so we don't create a problem. It's important to try to collaborate as much as possible," says Harvey. The uranium mines in Saskatchewan have worked hard to engage the Northern Saskatchewan residents in the business. The Industry contributes a stable, safe and relatively high paying work place for people from the area.

The team involved at ESO have devoted themselves to the hunt in northern Saskatchewan.

Ainsworth taps the core sample in front of him. "We're just coming to the most interesting part of the project because we're drill testing well defined targets. We've spent a lot of money on airborne geophysics and ground geophysics and we've built the background information database we need, now it gets exciting."

How exciting it will be for investors should start to become clear during the 2007/2008 drilling season. That could begin soon too, considering that ESO began a ten-hole diamond drill program on its 132,000 acre Hook Lake uranium property in early October. That operation, is being carried out using helicopter support to allow access prior to winter freeze-up, was indicative of Harvey and Ainsworth's eagerness to begin to "hone things down".

The Hook Lake drilling target selection was based on a combination of historical data and the latest technology – the combination that Harvey and Ainsworth believe must be used in this new wave of exploration in the Athabasca Basin. The targets are based on airborne and ground geophysical surveys, which included the high powered Fugro Megatem airborne system, high resolution aeromagnetic surveys and high resolution ground resistivity surveys. The drill target areas are near the up-ice limit of boulder trains with previously identified geochemical anomalies; these anomalies include boron, illitic clay, and other pathfinder elements that are seen in the alteration plumes above high-grade Athabasca type uranium deposits.

With about $6 million in cash and banker's acceptances with major Canadian banks and as advances to contractors, ESO has the financial muscle in its treasury to be aggressively exploring. ESO avoided the problems caused to some companies who were encouraged to acquire Asset Based Commercial Paper to boost interest returns on the capital raised for exploration. It is obvious that the ESO team can afford to manage every bit of its major land status in the Athabasca Basin with the selective contractor strategy it has chosen to follow.

ESO views itself as more than just another proximity play located beside a former producer and shoulder-to-shoulder with the most recent blockbuster discovery. The junior explorer has stand alone numbers of its own, highlighted by the historic CAR-425 hole from 1979 which showed results 350% higher than the global average resource quality for uranium mines. ESO also has an exploration team with experience moving from an idea towards a producing mine, and the financial and managerial know-how to see that through to a sustained operation.

For this junior explorer, the wild card for success will be time. All Harvey and the ESO team are missing is the needle pinpointing a sweet spot in the Athabasca Basin haystack. They are confident it is there and that they will be the next to announce a major discovery.

Forum Uranium Corporation

Mineral exploration has become much more sophisticated in recent years and any mineral explorer today operates an extremely complex business that requires meticulous planning. Adding to this complexity are the enormous risks involved in mineral exploration; these risks must be carefully assessed and managed if the venture is to be successful. Occasionally, a junior exploration company impresses more with its focused ability to manage this risk and run an exploration business.

One such company is Vancouver-based Forum Uranium Corp., an energy company that is exploring the Athabasca Basin of Northern Saskatchewan, and the Thelon Basin in Nunavut. The Athabasca Basin currently produces all of Canada's uranium, and the Thelon Basin has similar characteristics to the Athabasca with a major discovery at AREVA's Kiggavik-Sissons project.

Forum understands what is involved in the operation of a complicated business venture—the management team required, the need to attract investors who can supply capital, the use of new technology that conquers old challenges, and the risk management factors that must be included in a solid plan to mitigate what is often a strong gambling aspect of the sector. They perform SWOT analyses on themselves and their competitors—assessment of their strengths, weaknesses, opportunities, and threats. They use several analytical and decision making processes to determine the best way to proceed and then map out a disciplined execution plan once the project is under way. Forum knows itself well, clearly understands what they're getting into, and is cognizant of how to best operate once they're in.

Forum was incorporated in 1987 and was restructured as an energy venture in 2002 with the acquisition of a coal bed methane project. However, when President and CEO Rick Mazur joined the company in 2004 after a long career as a uranium explorer and senior executive in the exploration business, he recognized that given Forum's focus as an energy company, uranium exploration was something the company should consider. The market price for uranium was beginning to ascend at that time following the recognition that uranium would probably fuel much of the world's future energy production.

"I guess you could say I spearheaded Forum's move to uranium," Mazur says. "They were a long time explorer but were concentrating on coal bed methane. I could see early in the cycle that the uranium market was coming back and agreed to help Forum get the uranium business going. It was so early in the cycle that we were able to get a lot of good ground for a relatively low cost through claim staking. You can't do that any more."

Mazur, who graduated from the University of Toronto as a geologist in 1975 spent a decade exploring for uranium in just about every Province and Territory in Canada before returning to university for an MBA. After that he spent several years involved in business development and corporate development in the mining sector, where he gained experience in mine development, financing and investor relations. From 1985 to 1991 he was involved with the exploration, discovery, development, production, and eventual closure or sale of three Canadian gold mines operated by Canamax Resources. "At the end I was selling off scooptrams," Mazur recalls with a chuckle. "I really did go through the entire mining cycle. It was a very valuable experience that taught me all the aspects of mining."

Following the board's decision to move into uranium, Forum established a disciplined exploration and business development strategy aimed at minimizing the risks with any uranium exploration project. Because the company plans to eventually turn over any discoveries to expert mine developers, Forum also needed to ensure that accessing those discoveries would be cost-effective, thereby increasing the potential price of the properties. The company started with a sweeping analysis to determine where they would focus their uranium exploration efforts. The analysis considered where there was available ground with a high probability of uranium mineralization that was of sufficient grade, close to surface, and close enough to existing infrastructure such as transportation and processing facilities. The best fit for Forum was the Athabasca Basin, which is home to the world's largest and richest uranium deposits and which accounts for all of Canada's current uranium production. The Basin hosts almost 1.8 million pounds of uranium reserves, 620 million of which are at grades in excess of 20% uranium compared to the global average of only 0.24%. Given these high grades, the Athabasca Basin produces nearly one-third of the world's uranium and is regarded as one of the most profitable mining camps in the world.

The Basin itself is a 100,000 square-kilometre oval-shaped deposit of mid- Proterozoic sandstone that extends east to west and sits atop much older eroded Precambrian basement rocks that consist of highly-metamorphosed intrusive, sedimentary, and volcanic rocks. The contact point between the sandstone and the basement rock is known as an unconformity. The eastern rim area was explored by the major companies during the last uranium price spike in the late 1970s. While the prospects were subsequently abandoned when uranium prices slumped, this early phase of exploration yielded a great deal of geological data that is now publicly available and therefore readily accessible by today's smaller exploration companies.

All the known uranium deposits in the Athabasca Basin occur along structural corridors into basement rocks directly adjacent to these unconformities, which can be found as much as one thousand metres below the surface in the middle of the basin, or just a few metres below surface at the Basin's rim. Hydrothermal processes transported the uranium, which eventually pooled and crystallized in structural traps in the unconformities. But finding these structural traps can be difficult. In the basin, uranium deposits can sometimes occur as narrow, linear lenses at considerable depth.

Generally, such narrow targets represent significant drilling challenges. However, more advanced ground penetrating electromagnetic ("EM") geophysical surveying techniques help explorers track significant traces of graphite that frequently occur in the host rocks for the hydrothermal process that deposited the uranium. As a result, these signatures provide an excellent drilling map to the most promising zones of potential mineralization.

Because it was operating early in the uranium price rise environment, Forum was able to selectively choose available land in the Athabasca Basin that met its criterion- shallow targets nearby roads and uranium mills. Most mining activity in the area follows a trend line extending Northeast to Southwest from the Manitoba border just below the Northwest Territories. In fact, much of the Basin's known reserves and working mines such as Cameco's McArthur River mine are within the corridor. At the bottom of the corridor on the Eastern edge of the basin is the now shut down 200-million-pound Key Lake deposit and mine, discovered in the 1970s by Forum's Chief Geologist, Dr. Boen Tan for Uranerz Exploration and Mining Ltd., who recognized the geological conditions controlling the uranium mineralization at what was just a prospect at the time.

Forum's disciplined strategy when it moved into uranium exploration in 2004 was to look at existing uranium discoveries and determine how it could take advantage of its exploration expertise to discover more deposits along nearby trend lines. In the Athabasca Basin, the company determined that it would explore for uranium deposits along the trend line on the edge of the basin, where deposits were at their shallowest— as little as a few metres down— so that they might be readily accessible for a mine developer, and therefore be more cost-effective when it came time to develop the mine. Forum applied its knowledge and expertise to stake 1,600 square kilometres of prospective ground in the Athabasca basin, primarily on the northeastern and southwestern extensions of the corridor, along the edges of the basin. This ground includes a large package close to the crucial Key Lake Road area where characteristics similar to Tan's Key Lake discovery exist. A provincial highway through Forum's Key Lake Road claim gives it ready access to the Key Lake Mine's still-working ore processing mill—another cost-efficient factor that would make any subsequent discovery attractive to a mine developer.

The 100%-owned Key Lake Road project consists of 29 claims totalling 113,104 hectares and is located 20 kilometres southwest of Cameco's Key Lake Mine/Mill Complex, the principal processing facility for the nearby high grade McArthur River uranium mine and the site of the formerly productive Key Lake Deposit. Forum's permits cover favourable basement rocks within the Mudjatik-Wollaston Tectonic Zone, on the eastern rim of the Basin along which all current uranium producing mines in Canada are located.

Past exploration work in the Key Lake Road project area has been limited to surface trenching and shallow drilling of outcropping uranium showings and geophysical anomalies. Grab samples in outcrop located on newly staked claims graded from 0.032% to 7.65% U3O8. Past surveys have identified a number of untested electromagnetic conductors suggestive of graphitic zones in basement rocks within which enriched uranium mineralization frequently occurs. Further airborne surveys using higher resolution and more accurate helicopter-borne EM systems, such as VTEM and AeroTem, were conducted in 2005.

In September 2006, Forum embarked on a first phase program of shallow, widely-spaced drill holes to be completed during the 2006–2007 fall/winter drill season. Fifty drill holes on a number of targets in the Key Lake area were completed for a total of over 7,000 metres. Forum is planning an aggressive drill campaign in 2007/2008 to follow-up on results from the first phase of drilling and to test other new targets on the property.

The 100% owned Maurice Point project on the northwestern edge of the basin consists of fourteen staked mineral claims, totalling 37,714 hectares located immediately adjacent to Cameco Corporation's Maurice Bay deposit. The Maurice Bay uranium deposit is reported to host 1.3 million pounds of uranium at a grade of 0.6% U3O8 and includes structurally controlled mineralization within altered basement rocks and at the sub-Athabasca unconformity. The Forum claims cover extensions of basement structures containing the Maurice Bay mineralization and other structures along the favourable Athabasca unconformity.

Forum has conducted an airborne geophysical survey (GEOTEM) covering approximately 1300 line kilometres of the property. The GEOTEM survey was completed in May 2005. The data from this survey was compiled and combined with past geological and geophysical records and formed the basis of Forum's 2005 ground exploration programs at Maurice Point. Further prospecting encountered a similar style of mineralization as Maurice Bay at the Beach zone near McKenzie Point. The Beach zone uranium discovery and potential extensions based upon recent geological investigations have resulted in grab samples with reported assays from 0.3% to 7.3% U3O8. Further airborne magnetic surveys with the Goldak system have pinpointed drill targets that will be drill tested by incoming partner Mega Uranium Ltd. Mega will spend $8 million over three years to earn a 55% interest in the project.

Drilling results in both areas have convinced Forum that it was on the right track when it formed its extremely targeted plan. Mazur points out that Forum had to go against the conventional thinking to form its strategy, which is heavily reliant on Tan's and Forum's geophysical consultant, Phil Robertshaw's knowledge of the area during their tenure at Uranerz. Most explorers were still using old techniques that focused on drilling to unconformities that existed where the Proterozoic sandstone met the older eroded Precambrian basement rocks. But in 2000, the 50-million pound Millennium discovery was found by Cameco after it acquired the project from its takeover of Uranerz in 1998.

"Conventional thought was that uranium occurrences only existed at the interface between the sandstone and the basement rocks," Mazur explains. "But someone at Cameco's Millennium project just kept the drill going after Uranerz discovered significant alteration in the basement rocks and discovered Millennium. It was kept pretty quiet for five years, but it was finally publicized at a conference that I attended in Vienna in 2005. We had to really think far out of the box on this one but it worked, because we were able to acquire some really key ground along the trend in the eastern Athabasca for this style of basement deposit outside the Athabasca sandstone cover. Also, our Maurice Point property on the western edge of the basin sits in a similar trend line and we were able to acquire it by staking where similar shallow deposit potential exists."

"Since our business model was to focus on shallow targets, we were able to acquire another excellent project that met our investment criteria, the Henday property, at the top of the trend line next to AREVA's McClean Lake mill. It was explored for seven years by a private US company and we acquired the property for shares in Forum. We became aware of this acquisition opportunity as one person at the company was Tan's former boss at Uranerz, Klaus Lehnert-Thiel. A total of 32 drillholes have already identified some major alteration and uranium mineralization on the property and we plan to aggressively explore this prospect."

Forum's holdings in the Thelon Basin area, which is situated about 350 kilometres northeast of the Athabasca Basin in Nunavut, is geologically very similar to its more well known cousin in Saskatchewan. Still relatively unexplored, its similarities include the age and character of favourable host and basement rocks, major basement structural features, and known occurrences of unconformity-type uranium mineralization. One of the mineralized areas found to date is the Kiggavik Trend, host to approximately 130 million pounds of contained U3O8, which is currently being explored by AREVA Resources Canada Inc. AREVA holds the Kiggavik, Andrew Lake, and End deposits, collectively known as the Kiggavik-Sissons project. Other mineralized areas in the Thelon Basin include the Uravan/Cameco JV Boomerang Lake uranium–gold prospect on the western margin of the basin.

In July 2006, Forum and Superior Diamonds Inc. entered into a 50/50 joint venture, with Forum as operator, to explore for uranium in approximately 39,850 square kilometres of the northeast margin of the Thelon Basin near the Kiggavik-Sissons Project held by AREVA. The North Thelon Joint Venture claims are located southwest, north, and northeast of the Kiggavik-Sissons uranium deposits and occur in structurally uplifted areas of the same rocks hosting the Kiggavik-Sissons deposits. In late November 2006, Forum and its Joint Venture partner acquired an additional 34,892 hectares, expanding the total area in the Joint Venture to 101,562 hectares and also acquired an option to earn 60% on ground held by Tanqueray Resources Ltd. Evaluation of historical data in government assessment files list several uranium occurrences within the Joint Venture claims including the Shultz Lake and Boundary occurrences. The Shultz Lake occurrences are located within hematized and brecciated dirty quartzites (the same units hosting the Kiggavik-Sissons deposits) at two locations 100 metres apart. Assay results of rock samples returned values of 0.10% U3O8 and 0.31% U3O8.

Forum and its joint venture partner Superior undertook an aggressive exploration program in the summer of 2007 to further evaluate the significance of these and other occurrences. This program included ground geophysics, prospecting, geological mapping, and alteration studies of the Thelon sandstone in preparation for a drilling program.

The North Thelon Joint Venture is helped considerably by a new political climate in Nunavut that favours mining exploration and is the reverse of a former view held by Inuit leaders who opposed mining in the early 90's. In 2006, the Nunavut Tunngavik Inc., the body that manages Inuit-owned land, issued a new draft mining and exploration policy reversing its previous ban on uranium exploration and development on land it controls. With a view towards future economic development that would supply jobs and revenue, it also called for the adoption of a general approach to uranium mining on all Nunavut lands, not just that portion defined as Inuit-owned. The Uranium Mining Policy was ratified in September, 2007.

Mazur says that Forum is applying the same business model in Thelon as it uses in the Athabasca basin, primarily the exploration of the edges of the basin for shallow deposits and locating deposits nearby future infrastructure. "I spent four years in the barren lands exploring for uranium so I know the exploration history well," he explains. "I chose these areas because I knew we wanted to look at the shallow areas just as we are doing in Athabasca. And where there is one big deposit like Kiggavik, there could be more. With the recent addition of Ken Wheatley as our Vice President of Exploration to the team, his knowledge of the project will greatly increase our chances. Ken's last job was with AREVA, where he held the role of Regional Manager for the eastern Athabasca Basin, as well as project geologist for the Kiggavik- Sissons project. We feel that we have the expertise and knowledge to operate in Nunavut and will work with the local communities to the benefit of the Inuit people."

In fact, it is the expertise of this exploration team that allows Forum to be so disciplined in its thinking. Unlike many companies that merely stake around a previously discovered deposit in the hopes of finding some collateral deposits, often by sheer luck, Forum's plan is to increase its odds of success by employing the best exploration team in the business. Also, as evidence of its discipline, Forum sees itself strictly as an explorer, says Mazur. The company knows what it can do well, and so has no desire to develop or operate mines. Instead, it will leave that to firms that are expert in that area.

"We've developed one of the best teams in the country for exploring these kinds of deposits," he insists. "We obtained a lot of good ground by staking and permit applications relatively inexpensively. I've seen almost every geological environment for uranium in Canada and I believe that the Athabasca and the Thelon are the best by far. Our strategy is simple: we will assemble the best team possible that will look at shallow potential in likely areas where we know there's a good chance of making a discovery."

In some ways, the team that Forum assembled is a business planner's dream. It consists of several people who possess intimate knowledge of the Athabasca Basin and have been involved in several discoveries there. The team is headed by Rick Mazur, a geoscientist who has held positions in the international exploration and mining industry for over 30 years as a project geologist, financial analyst, and senior executive on uranium, gold, base metals, coal, and industrial mineral projects in North and South America. Mazur worked over a ten-year period as a uranium exploration geologist in Saskatchewan, Nunavut, the Yukon, Ontario, Quebec, and the Maritime provinces for Pan Ocean Oil Ltd. and then became an active consultant working for several mining companies including Canamax Resources Inc, Impact Silver Corp., Aurora Platinum Corp., Lakeshore Gold Corp. and Alto Ventures Ltd. He is currently a Director of Alto Ventures Ltd., Impact Silver Corp., and Tanqueray Resources Ltd.

Ken Wheatley, Vice President, Exploration, is a significant addition to the Forum team, Mazur feels. Wheatley is a Professional Geoscientist in Saskatchewan, Alberta, and the Northwest Territories/Nunavut. He has twenty-seven continuous years of uranium exploration experience in Canada, most recently over the last eight years with AREVA Resources Canada Inc., one of the world's largest uranium exploration and production companies. He worked at Uranerz before that where he got to know the rest of Forum's uranium exploration team. He has an impressive record of discovery including eight uranium deposits, four of which became producing mines in the Athabasca Basin. Wheatley also managed the Kiggavik-Sissons project for AREVA, a project that contains a 130-million pound uranium deposit in the Thelon Basin, where Forum has a strategic land position. "After leaving AREVA in 2007, Wheatley had fourteen different job offers but chose to join Forum", Mazur said. "He knew our Thelon and Athabsaca properties and liked what he saw, plus he liked the team we had. He thought the company had great value. And we have great value with him as our Vice President of Exploration."

Dr. Boen Tan, Chief Geologist, is famous throughout uranium mining circles for having discovered the Key Lake uranium deposit. A member of the Association of Professional Engineers and Geoscientists of Saskatchewan (APEGS), he possesses over thirty-five years of uranium exploration experience. He recently received the 2007 Outstanding Achievement Award from APEGS for his role in the discovery of Key Lake.

Tan joined Uranerz, a private German company in 1969 and after a number of years as a field geologist in Germany and Australia, moved to Canada in 1973 as a senior geologist and Project manager for Uranerz , conducting uranium exploration in the Athabasca Basin. Tan was instrumental in the discovery of the Key Lake uranium deposit and the development of the Key Lake Mine in 1976, which produced 195 million pounds of U3O8 at a grade of 2.5% over a fifteen-year mine life from 1983 to 1997. After the development of the Key Lake Mine, Tan continued to supervise Uranerz's uranium exploration and drilling programs in the Athabasca Basin, including regional exploration in the greater Key Lake area. Tan monitored the exploration and diamond drilling of Uranerz's joint ventures with Cameco until 1998. For the next seven years, Tan acted as a consultant to a number of companies within the uranium business and joined Forum as Chief Geologist in 2005.

"When I joined Forum in 2004, I dusted off my files on uranium, and discovered a paper Boen had written after discovering Key Lake," Mazur said. "He wrote the definitive paper on the unconformity- style deposit with Uranerz's Chief Geologist, Franz Dahlkamp. Phil Robertshaw set up a meeting for me with Tan in Calgary in early 2005 and we hit it off. Having Tan join our team significantly increases our chances of success because he knows every inch of this kind of ground and these kinds of formations."

Also on the team is Bruce Harmeson, Uranium Project Manager. Harmeson has been involved in the mineral exploration industry for thirty years. He was based in Indonesia from 1987 to 2003: for three years as a consulting geologist (gold) and twelve years involved in contract drilling services. Previously, he worked as a uranium exploration geologist in Canada and Niger for Pan Ocean Oil Ltd., where Harmeson and Mazur met as young geologists. Bruce also worked in northern

Saskatchewan with Uranerz , where he was a member of the Maurice Bay deposit discovery team. Bruce manages all logistics and permitting for the Company's projects.

Rounding out the expert Forum team is Philip Robertshaw, Geophysical Consultant. Robertshaw is a mineral exploration geophysicist with thirty-five years experience, a large part of which was gained in the Athabasca Basin area. His professional career includes six years as a geophysical contractor with a wide range of international experience and twenty years with Uranerz , based in Saskatoon, Saskatchewan, where he became Chief Geophysicist. Since 1998, he has worked as an independent consultant with clients active in exploration for uranium, gold, base metals, and diamonds. He is a member of the Society of Economic Geologists (SEG), the Geologic Association of Canada (GAC), the Canadian Institute of Mining (CIM), and the Prospectors and Developers Association of Canada (PDAC).

The company is in turn directed by a senior team of board members and advisors.

Anthony Balme, Director, is a finance expert who has been an active participant in a number of overseas resource ventures, both public and private. Balme is the Managing Director of Carter Capital Ltd. and A. M.C. Ltd., two private UK investment funds. He identifies early stage opportunities in the resource sector and arranges financings to advance projects to the public markets. Balme was one of the founding investors in Forum during its reorganization as an energy company in 2002. David Cowan, associate counsel with the law firm of Lang Michener LLP in Vancouver, is another of Forum's Directors. Cowan has practised in the area of securities and corporate finance for over fifteen years, and has represented numerous public and private companies in the mining and technology sectors. David brings a good measure of corporate governance to the Company.

A recent addition to the Board of Directors is Michael A. Steeves, CFA, who has been involved in the mining industry for over forty years. Steeves is currently the President and COO of Zazu Metals Corp., which is developing a large zinc–lead–silver deposit near the Red Dog mine in Northern Alaska. Prior to joining Zazu, Steeves was Vice President of Investor Relations for Glamis Gold, and also served as Director of Investor Relations for Coeur D'Alene Mines, Homestake Mining, and Pegasus Gold. Prior to his entry into the gold and silver business, Mike was a mining analyst with a number of investment banking firms in Toronto and Vancouver.

John Prochnau, a mining engineer and geologist, is a key member of Forum's Advisory Board. Prochnau has held management positions with several companies, leading to the discovery and development of the Alligator Ridge Gold Mine in Nevada and as a consultant, he has managed several private exploration syndicates and publicly-listed mineral companies responsible for a number of gold and base metal discoveries and mine developments in the United States, South America, and Australia. Mr. Prochnau was founder of Brancote Holdings PLC, whose Argentinean subsidiary Minera El Desquite S.A. discovered the high-grade Esquel Gold Deposit that was acquired by Meridian Gold Corp. for USD $368 million. As a result of these accomplishments, Prochnau has been inducted into the Kitco-Casey Explorers League.

Ed Schiller brings over 30 years experience in mineral exploration, project management, acquisitions, financing, joint venture negotiations and corporate governance to the Company's Advisory Board. He graduated with a degree in geology from Michigan State University in 1956, and obtained his Ph.D. in mineralogy at the University of Utah in 1963. Dr. Schiller was the Resident Geologist of the Northwest Territories Geological Survey of Canada from 1964-1966. He has lived and worked in Canada, the United States, England, Australia, Brazil, Columbia, and has conducted mineral exploration projects in several South and Central American, African and South East Asian countries, including Madagascar. Dr. Schiller has consulted for the United Nations on a gemstone project in Mozambique and a mining project in Greece. He has visited other countries on mining related projects, including Vietnam, Botswana and diamond mines in Yakutia, Russia and China.

Dr. Schiller is a former director of Dia Met Minerals Ltd., and is best known for supervising the drilling which led to discovery of the first diamond-bearing kimberlite at Point Lake, Northwest Territories in 1991, now part of the Ekati Mine production of BHP-Billiton. He has written extensively on the Lac de Gras diamond discoveries and has presented several papers on this subject at national and international meetings. Dr. Schiller writes for several Canadian and international magazines on mining and mineral exploration and maintains a consulting practice in British Columbia, Canada.

This kind of team building probably goes a long way to answering a lingering question indicated by Stockinterview.com in 2006. While analysing Forum, the investor education website asked simply "Will Lightening Strike A Third Time for Dr. Boen Tan?" It then went on to answer its own question by pointing out that the renowned exploration geologist was extremely familiar with Cameco Corp.'s Millennium deposit, which has been estimated to host 57 million pounds of uranium oxide, and is believed to be worth more than $2 billion. The geological setting of the Key Lake Road shear zone is quite similar to the Millennium deposit," Tan said. "It's also located in the same north-northeastern structural trend."

Tan's experience in the area and intimate knowledge of its geological characteristics accounts for Forum's elevation above the usual exploration method called "closeology",—the (sometimes deceptive) tendency for juniors to measure their property's value in relation to a major, often recently discovered, world-class deposit. "When the comparison comes from a highly regarded exploration geologist such as Boen Tan, one should pay attention," writer James Finch said. Apparently some did. A major shareholder in the company is Anglo Pacific Group PLC, a London, UK-based mining finance house which currently holds a 9.5% interest. Also, the company currently has $6.5 million in cash to finance its explorations.

Forum has an aggressive drill program planned for the coming year with 8,000 meters in the Key Lake Road area on its 100% owned Key Lake Road property, the Global Uranium option on the Orchid Lake property and the Costigan Lake Joint Venture with Breakwater Resources. That program runs from October 2007 to April 2008. In January, Forum plans to commence a 5,000 meter program on the Henday property and 3,000 meters on the Maurice Point property with earn-in partner Mega Uranium. . Due to the Company's near surface target model, close to 100 drillholes will be completed in the 2007/2008 field season. Between putting money in the ground and working on some very strategic corporate developments, Forum plans to have a busy and exciting year.

Unlike most junior explorers, Forum Uranium is not a risk-taker, although it certainly operates in a game that is risky by nature. Forum has entered the game well financed, studies its competition and is constantly calculating odds. Forum joined the uranium exploration game in 2004 with a focused strategy to conquer risks, stick to the areas it knew best, and explore the regions with the best potential for near surface deposits with existing and future infrastructure. Then it built a world-class exploration team to add to the chances its efforts would be rewarded. So far, says Mazur, "It's all coming together as planned."

Frontier Pacific Mining Corporation

Although Peter Tegart's office is simple and utilitarian, it does reveal something about the President and Chief Executive Officer of Frontier Pacific Mining Corporation. It's that simplicity that defines the man who heads a company that is currently engaged in developing mines in regions distant in not only geography but also in culture. The office, in one of the Bentall buildings in the heart of Downtown Vancouver, is functional and devoid of the flamboyant personal touches that are often features of the offices of other corporate chiefs. It has the standard desk, cupboards, and bookshelves of most offices, but they are unadorned and practical. The shelves are more likely to contain geological reports than flavour of the month business books; books that are not reports lean more toward cultural education than financial. For, example, a series of books on Mexico talks about the history, people, and culture of that country, not about its geology and mineral potential. Other books are about ranching and farming.

The overall impression is of a man who is in tune with himself and the earth, which is a good description of Tegart, an accomplished professional geologist with over 45 years experience in the exploration, development, and mining of precious- and base-metal deposits around the world. At the same time, it points to a man who is deeply in tune with the cultures of the regions in which he has operated or explored for mines. Most of these regions are remote and feature local populations of people who are also close to the earth. While working in these regions, Tegart brings with him a strong affinity and understanding of the nature of cultures in regions less developed than North America.

It's an understanding that comes from his own life, which has included poverty, farming, and most aspects of mining—from underground to the executive level. Tegart, 64, grew up in the Kootenay mountain and ranching region of British Columbia on a small farm that did not have running water, much less electricity and other "modern" accoutrements of civilization. He often rode a horse to a nearby schoolhouse and spent his summers working on his father's farm, where the family grew all its own food. Tegart fit the statistical definition of poor, but never felt that way. "We had all we needed," he explains, "we worked hard, had food and friends, and so didn't want for anything. I thought it was wonderful. I never considered myself poor."

Perhaps a testament to the joys of the ranching life is that Tegart today spends most of his summers on another farm in the same region, a hay-growing operation tucked into a mountain range just west of the Rockies and the Alberta border. It's a region where you're judged by your integrity and willingness to work, not by your wallet or the maker of your suit. "It's definitely work, what with all the chores that come with running a farm, but I like it," says Tegart, who still favours the simple garb—a golf shirt and khaki pants—of the working geologist despite more than a decade heading mining companies. "I like getting back to the kind of work where you can see what you've accomplished."

Tegart has had the mining bug since before he finished high school. Placer Dome operated a lead and zinc mine near his farm and at eighteen he left the farm to work machinery underground in the mine. After high school, he trained as a miner with CanEx for a few years to make some money to go to university. By now thoroughly ready to make mining his life's profession, he decided to study geology at the University of British Columbia and continued to work at various mines during his summer breaks. It was a heady time for the once poor Tegart who was now earning large incomes as an experienced miner. "I definitely wasn't a poor student," he said. "I was clearing more in a summer than most other people made in a year."

University training helped prepare him for the next stage of his mining career—exploration. Tegart joined El Paso Mining and Milling on a two-year prospecting contract and along the way became friends with legendary mining prospector Alan Kulan and Dr. Aro Aho, who were instrumental in discovering the giant lead/zinc mine in Faro, Yukon Territory the summer before with Dynasty Explorations. It was a friendship that was to have a lasting effect on Tegart. "I became interested in the mining business from a completely different point of view," he recalls. "I now saw it from the point of view of creativity. Those years with El Paso transformed me from someone who worked in mines to someone who discovered them."

In ensuing years, Tegart steadily climbed the ladder in mining exploration. For 15 years he worked in Canada, eventually as general manager, exploration, for BRGM (Bureau de Recherches Geologiques et Minieres), the French government firm that explores for minerals throughout the world. Among his achievements during this period was the discovery of the Lawyer's deposit in northern British Columbia. That deposit eventually yielded the Cheni gold mine, which Tegart raised capital for in the mid-1980s. By now the poor farm boy had become a company head responsible for working the capital markets in order to find and develop mines. But it was in 1990 that Tegart formed his largest company yet, Manhattan Minerals, which opened the Morris mine in Chihuahua state of northern Mexico. By 1998, the mine was producing 21,100 ounces annually, but severe drought conditions and a low gold price caused the company to suspend operations in 1999.

It was at Morris that Tegart began learning how to interact with different cultures. He had experience with Native Indian groups in Canada, but their culture and concerns were familiar to him. Mexico presented a completely different and more challenging scenario. The local indigenous people spoke different languages and tended to be wary of foreigners. The solution when working with the local population in a foreign region, Tegart believes, is to build trust through honest and continuous conversation. "You have to be interested in them as a people and be honest with them," he explains. "We talk to them in realistic terms because these are real people. I relate to them because of where I come from. We're weeds and seeds guys, basically earthy people, and I'm an old farm boy who doesn't like to bullshit." As he speaks about his interaction with locals, Tegart's accent unconsciously changes, dropping to a more noticeable folksy western Canadian twang liberally sprinkled with "eh"s. You can almost picture this head of a multi-million dollar mining company with a group of local villagers, squatting on his haunches and drawing pictures in the dirt as he explains what he proposes to do and how it will help them.

Tegart and Frontier Pacific are currently employing that methodology of trust-building as they develop a gold mine in northeastern Greece and explore for Uranium in southern Peru. The company has been in existence for 20 years—a moribund gold mining explorer that was reactivated in 2002 when Tegart joined with Mohan R. Vulimiri, current Director Chairman and Chief Geologist, to take advantage of rising gold and metal prices. Both current Frontier Pacific projects are potentially lucrative. The advanced stage Perama Hill gold property in Greece is expected to yield some 1.2 million ounces of gold, for a net present value of about $400 million at current gold prices and the Macusani uranium property is in the early drilling stages, but has the potential to be among the largest uranium mines in the world.

The Macusani Concessions are situated in the Province of Carabaya, Macusani District in south-eastern Peru. The concessions lie in the relatively flat Altiplano terrain at elevations between 4200 and 5000 metres above sea level. All of the uranium targets are accessible by gravel roads from the town of Macusani. Frontier Pacific formed a joint venture with Solex Resources in 2005 to acquire approximately 45,000 hectares in 67 concessions covering prospective uranium mineral occurrences in the district. The concessions have over 52 uranium showings and anomalies identified by the Peruvian Institute of Nuclear Energy ("IPEN").

Three major events of felsic volcanism have been identified within the Macusani area. The three felsic volcanic members, from oldest to youngest, are called the Chacacuniza, Zapanuta, and Yapamayo Members. Each geological member consists of a series of ignimbrite (massive and debris) flows disconformably overlying each other. The project area is bound to the north and east by older Proterozoic rocks. The Yapamayo Member dominates the eastern part of the project area and has been recorded up to 500 metres in thickness. Four different ignimbrite units have been identified within the Yapamayo Member. Each ignimbrite unit contains several facies varying from air-fall tuffs, waterlain tuffs, pumice-rich monomictic tuffs, polymictic tuffs and welded tuffs. Uranium mineralization is both structurally and stratigraphically controlled within a particular facies of two of the four ignimbrite units.

The near surface setting of the uranium mineralization potentially lends itself to very low mining and recovery costs for uranium. Initial agitation leach tests were carried out by Lakefield Research of Toronto under the supervision of Melis Engineering of Saskatoon to confirm the leachability of the uranium. The tests on the Sayana Prospect samples returned a high uranium extraction of 97.6% U_3O_8. Bottle roll leach testing is showing that high extraction rates of up to 89.8% of the contained uranium can be achieved in the first three days of leaching using mild acid solutions. The composite sample submitted for testing, with a uranium assay head grade of 3.8 pounds per ton (0.19% U_3O_8), is very low in associated deleterious elements. This would contribute to the ease of processing in terms of uranium extraction and recovery, capital costs, and downstream waste management.

Since late 2005, exploration on the Macusani Concessions has consisted of ground radiometric surveys over the known showings to determine the extent of the mineralization; detailed mapping and prospecting; channel and chip sampling of the showings; and channel sampling from ten trenches in an identified target area. Between April and June 2006, the Company completed 1780 metres of trial core drilling on four distinct uranium targets: Sayana Central, Calvario I, Puncopata, and Agaton.

Forty-four shallow holes were drilled from eight platforms in an area covering less than ten percent of the Macusani Concessions.

These four Uranium targets were identified by very large radiometric surveys completed by the Company in late 2005 and are separated by as much as 17 kilometres in distance. Drill results from four zones indicate good continuity and grade of uranium over substantial widths within coarse grained, predominantly flat lying, ignimbrite breccias. It is important to appreciate that, in each case, the trial drilling focused on a very small portion of the previously identified uranium bearing anomalies, which in many cases, measure up to several square kilometres in surface area. In December 2006, the 20,000 metre drill program began at Calvario III. This program has been designed to target the 7 prospects with the strongest geophysical anomalies from the exploration work carried out until November 2006. Four of those prospects were drilled in early 2006 and had significant uranium mineralization. A further three prospects were selected and included as part of the current 20,000 metre drill program. Two diamond drill rigs are currently working on the Macusani Project.

Tegart says Frontier Pacific became involved in uranium exploration almost by accident. The Macusani Concessions had been explored by the Peruvian government in the 1970s but the work was not continued because of a mining moratorium in the country, and because world uranium prices had sunk to $10 a pound in the wake of anti-nuclear sentiment. But the Peruvian government identified mining as an economic driver in the early days of the twenty-first century and in 2003 released the concessions to private companies. At the same time, the price of uranium ballooned. Solex Resources (TSX-V: SOL), a Vancouver-based exploration firm that works primarily in Peru and holds most of the Macusani property rights, approached Frontier Pacific about forming a joint venture to explore the holdings. Joint venturing is a risk-management strategy frequently used by junior explorations firms, who locate and explore new projects and then look for more senior partners to develop the mine.

"They came to Frontier Pacific because of our connections and reputation," says Tegart. "We have so much experience and so many contacts here. Vancouver is a powerhouse in the world of mining because it has been the generator of capital for discovery for a long time. But to generate that capital, you have to know what you're doing. You have to have the technical capability, but then you have to think about how you're going to do it, how you are going to generate investor interest. You have to work out the details: how widespread is the metal on the property? Is it consistent? Can it be mined economically? All those kind of things need to be worked out. If it has pretty good potential, that helps to convince seed investors, most of whom are based in Vancouver, that it's worth the risk."

Control of risk is Frontier Pacific's primary objective. In fact it's the basis of the company's business plan, which is simply to look for assets around the world that have most of the development risk taken out. That's why, as a mine developer that operates between initial prospecting and actual mine operation, it rejects most of the dozen or so proposals it receives every month. Generally, Frontier Pacific looks for small deposits with reasonably proven potential, which it then advances to the stage where it can hand over the developed property to a company that operates the mine. "We convert a promise to an asset," Tegart explains. "Seed investing has the highest risk because you're usually putting in your own dough, so we try to take a project to the next step, to where that risk is minimized. We can do it because we have the experience, the geological knowledge, and the credibility within both the mining and the financing communities."

Tegart points to his management team to back up his confidence in Frontier Pacific. The team is largely composed of extremely experienced mining and finance people. Perhaps, the most impressive among the Frontier Pacific team is Director Stewart L. Blusson, PhD, one of Canada's most successful exploration geologists. Dr. Blusson, who received his doctorate from the University of California (Berkeley) in 1964, discovered, with fellow geologist Chuck Fipke, Canada's first diamond deposit, now operating as the Ekati Mine in the Northwest Territories. It was a discovery which led to the creation of a Canadian diamond mining industry. His authoritative knowledge of Canadian and global geology led him to conclude that conditions for the occurrence of diamonds in Canada were favourable and accordingly he developed an exploration plan using highly refined scientific methods. Dr. Blusson is also president of Archon Minerals Ltd., a Vancouver-based diamond exploration company.

The team also features Mohan R. Vulimiri, PGeo, who acts as Director, Chairman and Chief Geologist. Mr Vulimiri received his Applied Geology degree in India and then earned a master's degree in Economic Geology from the University of Washington in 1975. A registered professional geologist and member of the Society of Economic Geologists and the Society of Mining Engineers, Mr. Vulimiri has 30 years of public company experience managing junior and established companies. His expertise is in the exploration and delineation of ore deposits with emphasis on structural controls and modes of occurrence of mineral zones. Mr. Vulimiri is also President of Nortec Ventures Corp.

Executive Vice President, Brian Lock, is an electrical engineer from England with over 30 years experience in the operations, feasibility, design, engineering, and construction of numerous precious- and base-metal mining projects around the world. Mr. Lock held senior positions with a major international mining consortium and two major engineering companies before forming his own company, Proton International Engineering Corp., a Canadian engineering and construction company devoted to the development of small and medium sized mining projects.

Another member of the team is Ian Laurent, Vice President of Exploration, who has a Master of Economic Geology degree from the University of Tasmania and 15 years experience in mineral exploration and development on four continents. Charged with helping Frontier Pacific advance projects to the bankable feasibility study stage, Mr. Laurent has actively explored in Guinea and Mali, West Africa, Turkey, and the Eastern Goldfields of Western Australia. From 1997 to 2000 Laurent was Project Geologist for Viking Gold Corp., responsible for designing and managing the exploration programs that discovered the Svartliden Gold Project in Sweden.

Another Director is Jonathan Goodman, MBA, CFA, PEng., who has 20 years experience in the resource and investment industry as a geologist, senior analyst, portfolio manager, and senior executive. Mr. Goodman graduated from the Colorado School of Mines as a Professional Engineer and holds a Masters degree in Business Administration from the University of Toronto. He is also a Chartered Financial Analyst and is the President and CEO of Dundee Precious Metals Inc., a Toronto-based mining company.

Tegart easily takes off his CEO hat to switch to his former role as an exploration geologist when explaining why Frontier Pacific's Macusani venture presents moderate risk to investors. When Solex approached it regarding the joint venture, the price of uranium had moved from $15 a pound to $30, a movement largely caused by the tremendous increase in interest generated by the worldwide energy and climate crises. As energy needs escalated because of rapid industrialization in regions such as China and India, and governments recognized that carbon release into the atmosphere by conventional energy production plants had to be curtailed, the world began turning once again to relatively clean nuclear energy production. Stalled since the 1970s when several accidents created an anti-nuclear hysteria, nuclear power was making a comeback as a cleaner, safer form of energy than conventional coal or oil.

"Even Greenpeace now prefers nuclear energy over carbon-releasing power plants, so there are something like 400 nuclear reactors being planned in the world." Tegart explains, "They require large amounts of Uranium 235, the active part of Uranium 238 that produces energy, but which makes up only 0.7 percent of uranium. Uranium consumption is 190 million pounds a year, but much of this is coming from nuclear warheads that are being de-commissioned by the US and Russia under their treaty. This supply will end soon and the next new uranium mine is years away. So the demand side is increasing and the supply side is decreasing. That's a big economic driver for the search for more uranium right now. At Macusani, there was a volcanic event that brought the uranium containing basement rock up and deposited it on the surface. It's kind of floating there and once we identify the best deposits, it can be recovered easily. Our story is simply that at Macusani, the uranium can be mined very economically. We think we can mine it for $10 a tonne, so we're looking at a return of more than six times gross metal value."

All business venture planning usually begins with a PESTEL analysis – the rather distant and refined study of the political, economic, social, technological, environmental, and legal situations and risks that might affect the venture. While Tegart understands this analysis, and can provide statistics to answer it, he has also learned that when you're dealing with people, sometimes even the best economics can not cover such things as social, political, and other risks. He comes by this knowledge painfully: He still bears the scars from the largest project of his life that went south because of these issues. That project was Manhattan Minerals' $240 million Tambo Grande project in an agricultural district of northern Peru that promised to develop a world class gold–silver–copper–zinc mine. In 1978, BRGM discovered mineral ore deposits in the subsoil beside and under the town of Tambogrande. After BRGM abandoned the project, Tegart's Manhattan Minerals, primarily funded by Canadian and American investment funds drawn to the area's potential, acquired its feasibility studies and 10,000 hectares of mining concessions. Manhattan Minerals Corp. undertook geophysical studies in 1997 and carried out exploratory drilling during the summer of 1999 (416 holes were drilled in nine regions). The results of exploration confirmed the existence of significant gold and silver deposits under part of the town of Tambogrande.

But the huge economic input into the area the mine would generate—$200 million (USD) in tax revenues and royalty payments collected by the government during the period the mine was in operation; 400 permanent jobs during the mine's existence; 1500 jobs during construction of the mine; 1000 jobs for the construction of new homes; and the offering of fifty percent of the jobs to the local population—were not sufficient to offset the impacts of having to relocate 8000 residents of the town. Despite spending $1,457,849 (USD) for some fifty socio-economic projects in Tambogrande and the nearby town of Piura—including literacy campaigns, the construction of wells, the purchase of a satellite dish and equipment for the Tambogrande television station, and funding of pre-university preparatory courses for hundreds of students in the region—Manhattan Minerals was unable to bring the community completely on side. Several Peruvian and international Non Governmental Operations (NGOs) successfully lobbied the townspeople heavily to fight the mine development.

Further, Peru's authoritarian and corruption-ridden government under President Alberto Fujimoro, who had awarded the mining rights to Manhattan, fell a year later. Successive transitional governments operated against a background of economic crisis, lack of confidence in public institutions, and decentralization of authority. Opposition to the mine development increased considerably and a municipal vote held in 2002 rejected the mining venture. Despite its best efforts, Manhattan had not won the hearts and minds of the local population, and had to abandon the project and its $70 million investment. It was a stark lesson in community relations that Tegart never forgot.

Armed with information learned from projects like Tambo Grande, Frontier Pacific is taking measured steps to avoid a similar situation in Macusani. For example, it has taken the measure of the local population's mindset and acted to please it. At the local elders' request, it has built two schools for local children and stocked it with several computers. "You have to build trust. It's how we do things now," Tegart explained. "And to build trust you have to try to fit your plans to the communities needs and honour your word. We can go in and think there are ten things they need, and nine of them won't be important to them. We were surprised about the computers, because it's a pretty basic society there with transportation being done on foot and with pack animals. They earn a living by growing potatoes and exporting them, along with llama wool. Computers were a quantum leap from the way they were living, but they're not stupid. They know what's coming and want their kids to be able to participate in the future."

Tegart's farm background allowed him to understand how the local farmers thought. He recognized that the uranium explorers were seen as intruders, foreigners who weren't part of the local community, and so took active steps to make his company become part of that community. All company representatives working in the area are Peruvian, speaking the local language—Quetchua—and the field manager is an ex-bullfighter–turned–geologist who was sourced locally and is fluent in the language. The company has also joined a national system that ensures half the tax revenue from a mine will come back to the district. On a more local level, company representatives meet regularly with elders to determine what the community needs and then provide it for them. This sometimes involves a canny understanding of local politics: Tegart points out that elders invariably ask for money first, because under the custom common to the area they hope to siphon off some of it for themselves. To ensure social benefits flow to the people of the region, Frontier Pacific asks instead what it can do for the community in the form of tangible buildings or infrastructure. Hence the computer outfitted school.

"You build trust by taking an interest in a people, by letting them participate in the operations. So we're offering a lot of local work. That way they see the impact quickly: there's now a payroll that provides the community with some money. You have to start with simple agreements and when those prove true, there's the beginning of trust. Then you go on to more complex agreements and build on that trust. It's participative and it's where mining is going. Mining companies are going to have to start offering this participation, the turning of local people into shareholders in the operation. It's the way things are going."

Frontier Pacific is using similar techniques in its Perama gold mining project in Greece, which it acquired from Newmont after local opposition blocked the development of the mine. From the year 2000 onward, local sentiment was extremely anti-mining, largely because there had been a large cyanide spill at a gold mine in neighbouring Romania. By 2004, when Frontier Pacific acquired the rights to Perama Hill, this anti-mining fervour had cooled somewhat although it still existed. Given the relatively poor economy in the Thrace region, largely based on tobacco farming which is now under threat from reduced EU subsidies, the Perama Hill project calls for considerable attention to be paid to community relations. The company has substantial plans for training, and up to eighty percent of its workforce will come from nearby towns and villages.

Also, the company launched a series of environmental studies and is embarking on an education program to point out facts and dispense information about the mine's benefits and safety. Five villages surround the mine, and Frontier Pacific's locally registered Greek subsidiary company is trying to visit every resident of each village to educate and argue the case for the mine. It's also bringing its message of environmental safety and sustainable jobs through public meetings, by visiting all social groups, by going door to door, by bringing local journalists to other mines to view how they are operated, and by composing a film about the project for the local television station. "You have to point out the economic benefits in real terms," Tegart says. "That gets people thinking and they start responding. It helps that I'm straightforward and talk with them. Now when I go there (Tegart visits the region regularly), I have three or four editors who always want to see me to discuss how the project is going. They recognize the mine will be a big economic help for the region that will supply many jobs, so they're much more enthusiastic."

In its Greek project, Frontier Pacific boldly states that "environmental and social responsibility is the most important component of the mine development...together with a social partnership with the people of Thrace." The same could be said for the Macusani uranium project, although it is still very much in the early and exploratory stages. The project could produce a much higher financial yield, estimated at about $3 billion, in future, but it's still a long road from exploration to active mining. However, Frontier Pacific is already applying a simpler version of its community relations methodology to form a "social partnership" with the local Quetchua-speaking farming community.

As Tegart indicated, any company that hopes to be successful in entering a rural area to develop a mine has to be aware of its status as an outsider, an interloper in what is very much a tight-knit society. Today, no mining company can arbitrarily land in a remote region and begin exploration and, inevitably, mining. The spread of democratic thinking and participative local politics, the rise of world wide NGOs and anti-corporate interest groups, and the ever watchful eyes of myriad environmental groups ensure that this past practice is long gone. Instead a company must proceed delicately to become, via a "social partnership," part of the community in which it hopes to operate. After simple trust is established, more complicated agreements that demonstrate "environmental and social responsibility" are required. With practice in some of the most far-flung mining regions in the world, there is no doubt at Frontier Pacific that they can pull it off.

Hathor Exploration Limited

Tucked in the northwest corner of Saskatchewan is a town so remote it's hard to imagine that anyone worked or lived there. But Uranium City was once a thriving mining community; often called "the last boom town." However, the uranium mines that drew people and enterprise into that remote northern town shut down in 1982.

Now, 25 years after the mines that supported Uranium City shut down, high economic hopes are once again pinned on this heavy, silvery-white metal. This time though it is the Athabasca Basin, a geological entity that straddles the Saskatchewan-Alberta border, that remains fertile ground for uranium prospects.; especially on the eastern side of the basin, where it sits on top of older rocks comprising the Wollaston Domain. Here, current uranium production accounts for close to one-quarter of the world's annual mine supply, making Saskatchewan the single-largest source of uranium in the world today.

According to the Saskatchewan government, the market value of the uranium extracted from Saskatchewan mines in 2006 was in excess of $600 million, and new mine construction is expected to employ 500 people a year for the next three years. It is estimated that over $1 billion is budgeted for the exploration for, development of and construction of the new uranium mines in the province.

Much of Saskatchewan's uranium is extracted from high-grade ore, the like of which is found nowhere else in the world. Ore considered high-grade in other uranium mining regions, like Australia, is often no more than 1/2% U_3O_8—high-grade in the Athabasca Basin is closer to 20% U_3O_8. Indeed, Saskatchewan's McArthur River mine frequently yields ore at such grades, earning it the distinction of being the richest in the world when it comes to in-situ value per tonne of ore reserves.

But the McArthur River deposit is small: it would fit within the area of a football field. That makes it easy to miss; especially at depths of 500 metres below the surface. Other deposits in the area are equally elusive. While Saskatchewan claims 12% of the world's known uranium resources, locating new deposits, proving them, and bringing them into production is no simple matter. Despite these challenges, the lure of great mineral riches has drawn company after company to the region.

Vancouver-based Hathor Exploration Limited. is one of those companies, competing to claim the next big discovery in the Athabasca Basin area. Armed with a strong enviable land position in the eastern Athabasca Basin and backed up by cutting-edge exploration technology, the company is among the strongest contenders seeking to make the next big uranium find in the region. Despite being a relative newcomer to the business, the expertise it's tapping as part of its current exploration program goes back to the days of the last boom.

In many ways, the story of Hathor begins with Dale Wallster. A veteran geologist who has made a career out of mineral prospecting across North America since 1979, he has focused his career on the exploration of metals he believes undervalued but have good potential for appreciation over the next three to five years. "I tend to look at a mineral commodity of significance," Wallster explains. "Then what I like to do is acquire a position in that commodity...basically to go out and acquire mineral exploration projects that ideally have a defined resource on them, but if they don't have a defined resource, then it's got to have the right geology and the potential to host a deposit of significance."

A case in point is uranium, which Wallster says he never expected to appreciate as strongly as it has. Recognizing the imbalance between existing supplies and growing demand for uranium, Wallster moved into the eastern Athabasca Basin in 2002. Early the following year he set up Roughrider Uranium Corp., a private company focused on exploration for unconformity-related uranium deposits. A seasoned prospector, Wallster steadily assembled and systematically explored promising land packages in the Athabasca Basin as the price of uranium rose from $8 (USD) a pound to over$100 (USD) per pound. At the same time, he forged relationships between Roughrider, Forum Uranium Corp. and Triex Minerals Corp., as well as other players with growing interests in the basin.

By the spring of 2006, Wallster had put together just over a million acres of properties, including several joint-ventured projects, giving Roughrider an exceptionally strong land position in the basin area. Roughrider's wholly-owned and joint-ventured properties ranged from the relatively small Midwest Northeast property to the large Russell Lake and Russell South properties—twenty-three claims covering 177,100 acres bordering the lucrative McArthur River mine.

These holdings were fair game when Hathor Exploration Limited came calling in the spring of 2006, looking to strengthen its position in the Athabasca region.

Roughrider had reached a point where it had to go public if it wanted to further capitalize on the properties it had amassed in the previous two years, and Hathor made it an offer that Wallster said "made a lot of sense."

Hathor had already staked its claim in the Athabasca Basin by acquiring six contiguous blocks in the northeast quadrant of the Carswell Dome—a site referred to as the Carswell property. A deal with Roughrider promised to augment Hathor's position in the area. A deal was cut that Hathor's newly appointed Chief Executive Officer, Stephen Stanley, believed would stand it in good stead for the future. "Roughrider is a strategic acquisition that offers Hathor shareholders a stake in one of the best available land packages in the Athabasca Basin," Stanley said in a statement regarding the acquisition when it was announced in mid-March 2006.

Hathor's acquisition of Roughrider closed in July of 2006, in an all-share transaction valued at just under $20 million. Today, repeating the familiar mantra of the real estate world, Stanley says the deal was largely about "location, location, location." The Roughrider properties formed a highly-prospective, diversified land package that gave Hathor 10 new projects as well as proximity to the infrastructure needed to process any uranium the claims produced. "The eastern part of the Athabasca Basin has infrastructure. You don't have that kind of infrastructure in the rest of Canada," Stanley says. "You have companies exploring for uranium in Quebec, in Labrador—but Cameco and AREVA's ore processing facilities in the Athabasca Basin are the closest uranium mills to any them, and they sit just a few kilometres away from our properties."

That is a huge advantage. While other companies have to make discoveries that will give them the tonnage and grade that would make mine development and the construction of a processing facility viable, Stanley contends that any discovery Hathor makes will be ideally positioned for development and thus more attractive to a potential mining industry major. He drives home his point, noting that current estimates suggest that the AREVA-operated McClean Lake ore processing facility, the most-modern in the world, would require a budget of $800 million to replace. "You need to have a significant deposit to justify that kind of plant," he says.

Even at the depths at which most of the uranium in the basin is located— almost a half- kilometer down—the high grades of uranium currently being mined in the region make production more viable here than in other jurisdictions. A shallow deposit that's easy to tap is good, Stanley admits, and if the price is high, even a low-grade, shallow ore may be economic. But he notes that if the price drops, the same won't be true. Declining prices can threaten the long-term viability of a project, something Hathor doesn't foresee in the Athabasca Basin. Ore from the McArthur River mine, for example, is of a grade that makes it economic even if uranium prices drop radically from their current high levels, something Hathor doesn't expect. "McArthur River, when uranium was at $60 a pound, had a net worth of $26 billion," Stanley says. "These things are absolutely incredible."

Moreover, McClean Lake isn't the only facility in the Athabasca Basin. There are two other mills serving the area. Throw in roads, access to power and other infrastructure required for a mining operation of any size, and the area is more appealing than other untested locations.

In addition to a significant land package in the world's number one uranium producing region, Hathor's acquisition of Roughrider handed it a significant bundle of expertise that gives it an edge over many other junior miners in the Athabasca Basin. Not only did the financial principals of Roughrider join Hathor, the technical expertise that had backed up Wallster's exploration efforts also joined Hathor. "To have retained his team of consultants, was a great accomplishment by Dale. Many other juniors, especially those coming late to the Athabasca Basin, have approached most of our people to try to get their services and they haven't been able to," Stanley says, clearly proud of the team Wallster developed and has helped Hathor retain.

The consultants Hathor retained through Roughrider have a much greater familiarity with the Athabasca Basin than most other consultants in the region. This is a team of people who stuck with the uranium industry through the 1980s and 1990s when the metal garnered very little investor interest. "They know the Athabasca Basin very well," Stanley says. "The great thing about these technical guys is that they've always been in uranium. So in a market like the one we've got now, where there's very little expertise in uranium, to have this kind of a technical crew is key." The advantage for Hathor is a better grasp on where it should be looking, what it should be looking for, and what it should be doing to improve its chances of making a discovery.

"Our team was right up to speed on what we should be doing in the field," Stanley says. "I think we've got an excellent shot at a discovery with our land package and the technical people we have."

But Stephen Stanley, even though he was only thirty-five years old when he took over as CEO, brought a lot of his own expertise to Hathor. At the time he took over, Hathor was just beginning to focus its attention on uranium among the several opportunities it had available to pursue. While the potential of uranium excites him, Stanley's expertise lies in management rather than exploration. What motivates him is managing a company properly, and creating value for shareholders. "I'm not necessarily motivated by uranium. It's not like it's a long-time passion of mine," Stanley says, speaking with characteristic candour. "I like putting good deals together. That's where I get my drive. Building the management team, raising the money, and moving things forward. At this point, a lot of those things have been accomplished, it's now time to drill and I'm excited about this opportunity."

Exploration and the opportunities it could reveal, are what excite Wallster and make him an ideal foil to Stanley when the pair go on the road to make presentations to shareholders and potential investors. Together, they give Hathor the drive and momentum it needs to be a viable and successful company. "Dale and I have actually worked quite well together," Stanley says. "I'm always looking to create as much shareholder value as possible. I'm always trying to stay focused on the share price, making sure that we're doing everything we can to benefit shareholders on that side, whereas he's passionate about uranium exploration. He wants to have the next major discovery in the Athabasca Basin. That builds a lot of investor confidence. They know we care about the share price and we're a legitimate company looking for a discovery." A discovery, Wallster adds, is where the big financial win will come for shareholders, making Stanley's life easier.

Stanley's expertise comes from what amounts to a lifetime in public companies. Growing up, he was surrounded by a milieu of public-company dealings through his uncle. During his twenties, Stanley was brought downtown and introduced to many of the players in the business by his uncle, relationships that eventually led him to move into corporate communications for a public company. "I got to know the industry, then I moved into the investor relations side," he says. "We started doing our own deals and it just evolved to where we are today. It's been very interesting. I've learned a lot, the hard way." During the intervening ten years, he's packed more than a decade of learning into his life through experiences heading his own companies (such as Sotet Capital, a privately held resource investment and capital consultant company) and serving as a director of other peoples' businesses, including KIT Resources Ltd (now Bayou Bend Petroleum Ltd.), and Aumega Discoveries Ltd (now

Fortress Base Metals Corp), both publicly traded companies listed on the TSX-Venture exchange.

Stanley began looking at uranium in 2003, just as prices were beginning to move north. "We thought it was a great direction to start heading," Stanley says. The result was that he formed a private exploration company which evolved into Titan Uranium Exploration Inc., with Arni Johannson in 2004 and began acquiring uranium assets in the Thelon Basin, an area in Nunavut considered to be geologically similar to the Athabasca Basin. Johansson has since taken the company public on the TSX-Venture exchange, and acquired properties in the Athabasca Basin as well, following the stampede that has snapped up most of the prospecting sites in the area.

When Stanley moved to Hathor early in 2006, Hathor's holdings in the Athabasca Basin were limited to the Carswell property. It also had a half-interest in a permit to 19.2 million acres as well as a full interest in 600,000 acres of prospective uranium properties in the Northwest Territories. While the area hasn't yielded much to date, it did give Hathor bragging rights as the largest landholder in the area prior to Cameco moving in, which was a boon in itself—it ranked Hathor among the pioneers in an area that is starting to see major interest from the industry's larger players.

The company also has a large, 300,000-acre package in B.C., the Eskay Creek lands that show some promise for gold and silver as well as base metals. But as Hathor focuses on its uranium prospects, Stanley plans to spin the Eskay Creek properties off into a separate company. The proximity of the property to Barrick Gold Corp.'s own productive holdings in the area make it something worth holding on to.

The streamlining of Hathor's holdings position it for advances in the race to discover a new uranium deposit almost as much as the key acquisition of Roughrider did by giving the company the raw material it needed to pursue that course.

Since Stanley took the reins of Hathor from former president and CEO Matthew Mason in February 2006, he has not only developed its property holdings but he's taken steps to strengthen Hathor's management and advisory teams. Of course, there was already good expertise available when Stanley came on board. Former president and CEO Matthew Mason, now a director of the company, was a long-established, self-employed businessman and investor with a background in corporate governance. He also had experience as a director and officer of several companies listed on the Toronto Stock Exchange as well as the TSX-V. For several years, Mason had been a senior executive or director of many public companies in the business of natural resources exploration and development.

Hathor also counted engineer Benjamin Ainsworth among its Directors. A senior geologist and mining consultant whose work in the mining industry stretches back over thirty-five years, Ainsworth had studied at Oxford University in England. He joined Placer Development in 1965 and held positions of Senior Geologist, Chief Geochemist, Exploration Manager (Eastern Canada), Exploration Manager (Chile), and served as Placer's President in Chile and South America for several years. Throughout the 1970s, Ainsworth was involved in the design, budgeting, and implementation of exploration programs that included large and small drill programs, geophysical surveys, geological mapping, geochemical surveys, and a full range of project evaluation studies.

To that line-up, Stanley has overseen the addition of some attention grabbing financial and management talent. John Currie, formerly Chief Financial Officer of resort developer and operator Intrawest Corp., came on board shortly before Intrawest's takeover in late 2006. He is currently CFO of Lululemon Athletica, Inc., and steered that company through its ambitious IPO during the market turmoil of 2007. "John brings a lot of wisdom. He's not a mining or an exploration person, but that's not why we have him on board. We have him for his business expertise," Stanley says. But Currie was also enticed by the prospects for uranium. "He was excited about the future of uranium," Stanley says, bluntly.

During the second half of 2007, Hathor was able to attract former Chief Executive Officer and President of Bank USA and HSBC Bank Canada, Martin Glynn, to its board of directors. A member of the Board and Audit Committee of Husky Energy Inc., Glynn's international experience lends a broad and ambitious perspective to Hathor's decision-making abilities "He's been a great asset to us," Stanley says.

Also in the second half of 2007, former Intrawest Playground CFO Andriyko Herchak joined as CFO of Hathor, courtesy of John Currie. He's also young, about Stanley's age, and lends more youthful dynamism to the organization. "We're both very keen on working hard and having success, so I'm very pleased to have him join the company," Stanley says. All in all, the appointments of Currie, Glynn, and Herchak are the type Stanley believes will stand Hathor in good stead as it struggles with the difficult decisions a junior mining company often has to make. "We've done well with bringing some very smart people on board," Stanley says.

As Hathor pushes forward with aggressive drilling programs, the expertise will be needed to ensure it stays on track and makes the best use of the more than $20 million it has in the bank. While Stanley expects Hathor's burn rate to be moderate, with the cash supporting more than two years' worth of exploration, prospecting activities aren't all the company is about. "We try to put 90 per cent of the capital we raise into the ground," Stanley explains. "But we must also focus a portion of our capital into creating investor awareness, there needs to be a balance". You can be doing all the right things but if investors are not informed your market will suffer!"

While Hathor's stock price has surged from less than 50 cents to more than $2.34 since February 2006, Stanley is not satisfied, he knows they have to stay focused on achieving the company's goal of finding a worthwhile uranium deposit. "I'm looking for that drill hole that absolutely lights up this company," he said.

To that end, Hathor has plans for advancing its properties. "What we want to do is hit our projects with major drill programs," he explains. "Investors don't have a long attention span. They want it now, and so the typical junior will go in and they'll drill three to five holes and come up with nothing and scratch their heads. We're trying to hit our drill targets with ten to fifteen drill holes as a first pass." Drill programs of this size will give Hathor's geologists the data they need to better understand these properties, and increase the chances of hitting that first great intersection.

Having the help of the technical consultants recruited under Dale Wallster will help the success of the drilling program. Based out of Saskatoon, a city that has earned its reputation as the scientific centre of the province, the consultants bring backgrounds with the Saskatchewan Research Council and the University of Saskatchewan's Depatment of Geological Sciences to Hathor as well as the connections that enable the company to tap into leading-edge technology for the identification of promising deposits. One such process being employed by Hathor to help locate ore deposits is a combination of the physical and the digital— seismic tests that create three-dimensional images of what's under ground . The process tests for structure, and identifies promising zones to drill. What the imaging is helping to identify is what's known as the unconformity—an area where the Athabasca Basin sandstones meet the older basement rocks. "Hathor and its technical team believes very strongly you cannot have a uranium deposit unless there's structure," Stanley said. "So we look for deep-seated structures that come up from the basement, we then look for displacements at the unconformity. Where the basement structures cut across the unconformity; that is a priority drill target."

The images generated are analyzed by a technical team at the University of Saskatchewan in Saskatoon, which has access to a special 3D imaging and analysis centre. The results to date have impressed even the majors. "They were actually quite blown away by what Hathor is doing," Stanley said. "That's what makes us very much cutting edge, and when people see the work that we're doing, they know we are serious about a discovery." Stanley asserts that the efforts of many other junior resource companies amount to a long-shot without a disciplined, technologically advanced approach. "You can have an address, in the Athabasca Basin and I guess that gives you a chance of finding something," he says, "but without the right ground, the right tools and the right people it's very much a long shot."

Through the winter of 2008 Hathor expects to use its findings from the 3D analysis to target its drilling on the Midwest Northeast claims. While the challenge of finding labour will pose an obstacle, Stanley has an aggressive drill program to meet and is eager to see the results.

Hathor has five projects where it will be actively drilling in-2007 and 2008—Russell Lake, Russell South, Wollaston NE, Carswell, and Midwest Northeast. It will be the busiest time for the company since its presence in the Athabasca Basin began.

Progress to date is already encouraging, with several projects showing promise.

Milliken Creek, for example, a 3,995 hectare property in which Hathor owns a 100% interest, lies only 80 to 125 metres above the Athabasca unconformity, just 26 kilometres from Cigar Lake, where proven and probable ore reserves are grading at 19.06% U_3O_8. Hathor's joint-venture with Triex Minerals at Old Fort Bay, a project covering 87,040 hectares is approximately 35 kilometers from AREVA's formerly-producing Cluff Lake mine and about 50 kilometres northeast of AREVA's Maybelle River property, where drill intercepts of up to 21% U_3O_8 over five metres and individual assays of up to 54.5% uranium, or 64.27% e U_3O_8. The Haultain River and Vedette Lake projects are both southwest of past-producing mine operations at Key Lake, once the world's largest supplier of high-grade uranium. Indeed, in 1997 the mine was providing 15% of the world's uranium.

Of course, most encouraging for Hathor's hopes at the moment are the properties it holds that immediately border mines, whether currently in production or past producers of note. Hathor's Russell South and Russell Lake projects, packages that comprise a total of approximately 71,670 hectares, border the McArthur River mine property of Cameco and AREVA where proven and probable ore reserves average about 20% U3O8. The projects also border the Moore Lake property of the new Denison Mines and the Wheeler River property of Denison and Cameco, where drilling continues, and AREVA and Cameco's formerly producing Key Lake mine.

But the most promising for Hathor to date has been the relatively tiny Midwest Northeast project, acquired through its 2006 deal with Roughrider. It holds a 90% interest in the 502-hectare property, which is located approximately 4 kilometres northeast of the Midwest uranium ore body owned by AREVA, Denison Mines Inc., and OURD Canada Co. Ltd. The Midwest ore body grades 5.47% U3O8, 4.37% nickel and 0.33% cobalt. It is also just 900 metres northeast of the Mae zone discovery where ore grading at upwards of 26.7% U3O8 has been identified. Denison, for its part, has reported recent drill results of up to 15.3% U3O8. AREVA considers the Mae zone to be the best uranium discovery in the Athabasca Basin, and possibly the world, in the past three years. Moreover, the shallow depth of approximately 200 metres allows for open pit mining and makes the prospects for any discovery particularly exciting. Hathor will commence its first major drill program on the property in the winter of 2008, following up a 3D-seismic survey, and hopes to complete more than 20 drill holes by the end of April 2008. "To be able to look down and see a conductor that's responsible for the Midwest deposit and the Mae zone continue right on to your ground is a pretty exciting thing," Stanley says.

The good news is, it's not just Hathor that's been impressed with its progress. In June 2007, Hathor was selected as one of the "2007 TSX Venture 50," a designation that recognized the company as one of the top mining stocks trading on the exchange. Only two other uranium companies were chosen, including Vancouver's UrAsia Energy Ltd. (acquired by Uranium One Inc.) and Boucherville, Quebec-based Strateco Resources Inc. The designation also acknowledged Hathor's top-tier standing out of 2080 potential TSX Venture Exchange-listed corporations that could have received the honour. The award is based on a ranking formula with equal weight given to revenue in the previous twelve months, return on investment, market cap growth, and trading volume.

Uranium
Corporate
Snapshots

Adriana Resources Inc.

Stock symbol: ADI – TSX

Rick Barclay and Mike Beley at the corporate offices of Adriana Resources in Vancouver

Down hole gamma logging for uranium at Tabb Lake area, Nunavut

Mcgregor Lake Camp at MIE project in Nanavut

Tabb Lake, Nunavut historical high grade uranium occurrences are being confirmed by the field crew

701 W. Georgia Street, Suite 1818, Vancouver, BC V7Y 1K8
Phone: 1.604.629.0250 • Website: adrianaresources.com

Alberta Star Development Corp.

Symbol: ASX – TSX

Gilbert Labine, discoverer of the Eldorado Mine

K-2 Zone drilling

President and CEO Tim Coupland inspecting poly-metallicmineralization at Contact Lake

Uranium outcrop with copper, zinc, nickel and cobalt

675 West Hastings Street, Suite 506, Vancouver, BC V6B 1N2
1.604.681.3131 • alberta-star.com

Cash Minerals Ltd.

Symbol: CHX – TSX

Drill core - magnetite, breccia

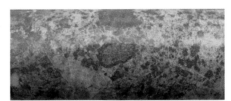

Drilling at Igor

Helicopter bringing in supplies

Suraj Ahuja and Dr. Geordie Mark discuss structural drilling

1066 W. Hastings Street, Suite 1890, Vancouver, BC V6E 3X1
1.604.633.9942 • cashminerals.com

Dejour Enterprises

Basin

Bolton Airstrip

Bolton Base Camp

Drill Core

808 West Hastings Street, Suite 1100, Vancouver, BC V6C 2X4
1.604.638.5050 • dejour.com

ESO Uranium Corporation

Chairman Tony Harvey, VP Exploration Ben Ainsworth and Corporate Communications Manager Tom Corcoran (left to right)

ESO Uranium's land holdings throughout the Athabasca Basin

= ESO URANIUM's Athabasca Basin Properties.   Totalling 985,590 acres

Inspecting assays from the Bridle Lake zone drilling at the Eso core shack on the Cluff Lake property

Oxidized uranium from Cluff Lake

1199 West Pender Street, Suite 408, Vancouver, BC V6E 2R1
1.604.629.0293 • esouranium.com

Forum Uranium Corporation

Symbol: FDC – TSX

Bob Nichol, Boen Tan, Rick Mazur and Anthony Balme at the Key Lake Road project

Jacques Stacey, Ken Wheatley and Boen Tan at the North Thelon JV Project

Scintilometer reading Key Lake Road Project

Map of key properties

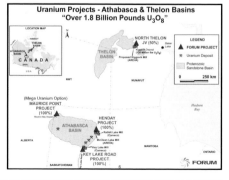

Uranium Projects - Athabasca & Thelon Basins
"Over 1.8 Billion Pounds U₃O₈"

475 Howe Street, Suite 910, Vancouver, BC V6C 2B3
1.604.689.2599 • forumuranium.com

Frontier Pacific Mining Corporation

Symbol: FRP – TSX

*Community Communication
in Isivilla*

*Drilling on Calvario I looking
at Puncopata*

*Exploration Team reviewing
Prospect location from Condorillo and
looking south towards LaHuana 1*

Peru Exploration Project Location

555 Burrard Street, Suite 875, Vancouver, BC V7X 1M8
1.604.717.6488 • frontierpacific.com

Hathor Exploration Limited

Symbol: HAT – TSX

Leading the way in High-Tech Exploration

Location, Location, Location

Hathor's Athabasca Basic Projects

3D Virtual Reality Centre - Saskatchewan Research Council

925 West Georgia Street, Suite 1810, Vancouver, BC V6C 3L2
1.604.684.6707 • hathor.ca

International Enexco

Symbol: IEC – TSX

Arnold Armstrong, CEO
International Enexco

Contact drill

Mann Lake property map

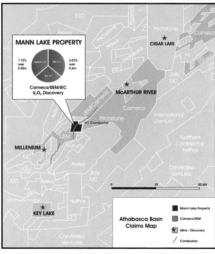

Moving drill to new location

777 Hornby Street, Suite 2080, Vancouver, BC V6Z 1S4
1.604.633.4280 • enexco.ca

JNR Resources Inc.

Symbol: JNN – TSX

Examining Core At Moore Lake

Massive Pitchblende vein, Hook Lake showing0482

More Lake Camp

Way Lake Core

315 – 22nd Street E., Suite 204, Saskatoon, SK S7K 0G6
1.306.382.2211 • jnrresources.com

Mawson Resources Ltd.

Symbol: MAW – TSX

(l to r) Andrew Browne, Folke Soderstrom, Mark Saxon, Klappibacken Uranium project May 2006

Consulting geologist John Nebocat (left) and David Henstridge, Sweden 2007

Mark Saxon Swedish Caledonides 2006

Michael Hudson with geopick and scintillometer - Spain 2006

1090 West Georgia Street, Suite 1305, Vancouver, BC V6E 3V7
1.604.685.9316 • awsonresources.com

Pan African Mining Corp.

Stock symbol: PAF – TSX

CEO Irwin Olian with new friends

Deep core drilling

Tranomoro base camp

Uranothorianite

650 West Georgia Street, Suite 1925, Vancouver, BC V6B 4N8
1.604.899.0100   panafrican.com

Silver Spruce Resources Inc.

Symbol: SSE – TSX

A chopper's eye view of the camp

A picture at the 649 Find

Core sample classification

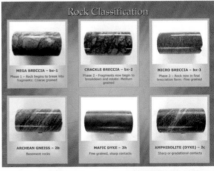

*President Lloyd Hillier, Prospector
Alex Turpin, and Senior Geologist
Guy MacGillivray*

102 Aberdeen Road, Bridgewater, NS B4V 2S8
1.902.527.5700 • silverspruceresources.com

Strathmore Minerals Corp.

Symbol: STM – TSX

Monitor well drilling at Gas Hills, Wyoming-George Ver (October, 2007)

Monitor well drilling at Roca Honda, New Mexico (July, 2007)

Strathmore management team at Sumitomo head office Japan (l to r) Dev Randhawa, Dieter Krewdl, Ray Larson, John DeJoia, David Miller

Strathmore Minerals Corp

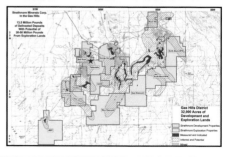

1620 Dickson Avenue, Suite 700, Kelowna, BC V1Y 9Y2
1.250.868.8445 • strathmoreminerals.com

Titan Uranium Inc.

Symbol: TUE – TSX

*Diamond drilling in the
Athabasca Basin*

*Exploring the geological structures
of the Thelon Basin on
Nunavut properties*

*Helicopter and drill on Titan's
Thelon Basin properties in Nunavut*

*Philip E. Olson, Chairman, CEO
and Director of Titan Uranium*

2100 Airport Road, Suite 100, Saskatoon, SK S7L 6M6
1.306.651.2405 • titanuranium.com

Tournigan Gold Corporation

BorisMikeJimLatsoOther

Drill positioning at the Kreminca gold deposit in Slovakia

John Cuthill and an associate, logging core from the Curraghinalt deposit in Ireland.

Logging core at the Kuriskova deposit in Slovakia.

570 Granville Street, 12th Floor, Vancouver, BC V6C 3P1
1.604.683.8320 • tournigan.com

Trigon Uranium Corporation

Symbol: TEL – TSX

Drilling at the Henry Mountain

Henry Mountain drill site

Sid Himmel in mine in Utah

Stu Havenstrite, Technical Advisor and Magnus Haglun, COO at Henry Mountain

1889 Spall Road, Suite 203, Kelowna, BC V1Y 4R2
1.250.763.5533 • trigonexploration.com

Uranium One

Symbol: UUU – TSX

Drilling underway at Kharasan

Incline shaft and decline access to orebody

Ion exchange columns

Wellfield control station

390 Bay Street, Suite 1610, Toronto, ON M5H 2Y2
1.416.350.3657 • uranium1.com

Ur-Energy Inc.

Symbol: URE – TSX

2007 Lost Creek Drill

Lost Creek water supply well

New geological logging truck

Wyoming Sky

10758 W. Centennial Road, Suite 200, Littleton, CO 80127

ur-energy.com

Waseco Resources Inc.

Symbol: WRI – TSX

Field airborn geophysics

Geophysical map of Block 1

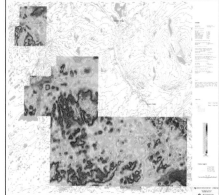

Panoramic view of Block I

Panoramic view of Block III

55 University Avenue, Suite 502, Toronto, ON
1.416.364.3123 • wasecoresources.com

Preceding this recognition by the TSX-V, Hathor had received kudos from U.S. stock-watcher James Winston, who in his October 2006 newsletter, delivered effusive praise of Hathor's prospects based on the results of 2006 survey and drilling activities. Praising the Athabasca Basin generally, Winston particularly noted Hathor's Russell South and Russell Lake projects for being in one of the likeliest areas of the basin to report a new discovery. Winston also singled out the Midwest Northeast project for compliments in light of findings in the Mae zone. The two strongest results from Denison Mines' drilling were closest to the Hathor property, boosting chances that Hathor may be sitting on a major discovery. "What's important about their properties is that they cover the right geology—the geology that currently accounts for about 30% of the world's annual mine production of uranium," Winston concluded. "This fact alone makes Hathor a much better stock selection then most of the companies out there." And for Winston, the expertise the company has amassed gives it a good a shot at success. "As far as grassroots uranium exploration companies go, you'll be hard pressed to find one with such well-positioned properties and probably the best geology team you'll find outside of a blue-chip company," he said. "If they are successful at finding high-grade on any one of their properties, this stock will be a ten-bagger."

But Winston also asked the question everyone—including Stanley—has been asking of Hathor: "Do they have the goods?" The question rings in Stanley's ears not just because of his own eagerness or the curiosity of pundits like Winston, but because of the typical impatience of shareholders. Savvy to investors' desires to see results, especially during periods of market turmoil, Stanley knows drilling over the next 18 months has to tell the tale. "Our shareholders have had to be more patient with us than with other juniors, they are always asking, 'When are you drilling? When are you drilling?'" Stanley says. It sure feels good to finally not only be able to say we are now drilling, but to be able to say we are now drilling very high quality drill targets and we are hitting them hard with major programs!"

Well, drilling is underway, and a discovery is—he believes—waiting to happen. When it does, shareholders will have what they want, and then some. There will also be decisions to make about what to do with the offers that are bound to come flooding in. "It's any company in the basin," Stanley says. "It does not matter what junior, if you make a discovery, you'll most likely become one of the hottest uranium stocks and you would then have the majors seriously looking at you. These deposits are very small and with Athabasca type of grades it does not take a lot of ore to make something economic."

In the meantime, Stanley remains focused. He knows the risks, and reminds himself of a discussion he had with one of Hathor's geologists. "The analogy that's used is, lay a shoestring on the floor and put a few raisins down on it, drop a pencil and if you hit the raisin you win. We know we have the structures (shoestring) on our properties and once you find that, you're on the right track," he said. "Is there uranium associated with the structure? We don't know yet. but what we've got here is as good as it gets right now.'" And if it gets better, it won't be by the number of holes drilled, but by the single hole that yields a discovery worth taking into production.

"We're really after one hole," Stanley says.

International Enexco

When Arnold Armstrong looks out of his 20th floor boardroom window he has a sweeping floor to ceiling view of downtown Vancouver. He can see the distinctive copper roof of the Hotel Vancouver, Burrard Inlet, and the North Shore Mountains framed in the distance. On one wall a Richard Tetrault painting shows an abstract view of the Port of Vancouver, and another wall is lined with law books. One can imagine Armstrong contemplating the huge changes that have taken place in Vancouver over the last five decades and reflecting on the role he has played among them.

More than half a century after being called to the British Columbia bar, Armstrong continues to practice corporate and commercial law as a barrister and solicitor with Armstrong Simpson. Of his four children, two of his sons are also lawyers; Brad is on the board of International Enexco and Mike practices law at Armstrong Simpson. Mike's specialty is defending other lawyers charged with negligence. Now 81 years of age, though looking and sounding many years younger, Armstrong heads up Armstrong Simpson, a law firm that acts for security firms, mining companies, and oil and gas companies. The law offices are also the headquarters of International Enexco, the exploration company that Armstrong heads as President and Chief Executive Officer.

Armstrong made his first serious investment in the resource industry in the early 1960s. The young lawyer put $12,000 into a venture company and promptly lost it all. It was an expensive lesson, and it taught him that he should be on the other side of the table. Not wasting any time, Armstrong set up a company called Christina Lake Mines and went into the mining business. From Christina Lake Mines it was a short jump to Pyramid Mines, one of the all time success stories of the industry. It was Pyramid Mines that discovered the massive lead–zinc deposit at Pine Point in the Northwest Territories in 1965. "We were lucky enough to hit a substantial deposit, the stock went up to $22 and there was pandemonium on the streets," explains Armstrong. "It was a real coup back in those days." Armstrong started by buying-up fifteen claims, and by the time he was finished, he'd either staked or bought-up 427 claims. According to newspaper headlines of the time, the stock kicked off a wild buying spree and put the Vancouver Stock Exchange on the world map. "It was just unbelievable," recalls Armstrong. "There were people stopping their cars right in the middle of Howe Street and dashing into the Vancouver Stock Exchange and they'd say buy me 1,000 shares of anything. A mail man who owned 6,000 shares dropped his bag and said "I quit!" and off he went—all of a sudden he had $120,000." In fact, Pyramid gave the VSE its biggest day in trading history and set off a flood of investment into the entire resource industry and a rush to stake claims. "Thousands of claims were staked right down to the British Columbia and Alberta border," he says. Pyramid Mines sold the claims to Pine Point Mines for $33 million in 1969.

Ironically, Armstrong found the only other lead-zinc ore body in the area through a claim he'd staked with another company. It was only about a tenth of the first windfall, but he sold that one as well to Pine Point Mines for $1.2 million. It was a huge win for Armstrong and gave him instant credibility in an extremely fickle industry. "There's an old saying in the industry," says Armstrong. "If you are speculating you have to make it on the first one and then you can lose on the next couple. But if you are a risk taker, it all has to be calculated risk."

And, no question Armstrong is a risk taker. While it's pretty hard to top a major discovery like Pyramid Mines, Armstrong managed to follow that up with a string of successes. After selling the company he jumped into the oil and gas sector. He formed a company called Coralta Resources, drilled a number of holes in Strathmore, Alberta, put in several miles of pipelines and a compressor station and built it up to where it had a cash flow of $50,000 a month. Armstrong still retains a 25% interest in the property through View Mont Estates, a real estate development company. Over the years he has held an interest in Ivory Oil & Minerals, and in UGL Enterprises, now called Red Hill Energy which holds uranium and coal properties in Mongolia. Armstrong is past chair and CEO of the remarkably successful SKN Resources, now known as Silvercorp, a hugely profitable silver producer in China. In fact, it was Armstrong that brought in Dr. Rui Feng who discovered the world-class Ying silver deposit in China and made Silvercorp one of largest silver producing companies in the world.

Being a lawyer, he says, has been incredibly helpful in the mining exploration business. "Your legal training teaches you how to reason things out and how to examine it from every aspect so you really reduce the risk because you anticipate what could occur and you move forward from there," he explains. It also helps with one of the most difficult challenges in the industry—raising venture capital for exploration. "The most difficult time was raising money. First of all you had to go to the brokers and they had to have confidence in you, because if they don't they are not going to sell it to the public," he says. "Then when you raise the money you have to treat it as though it's your own. And it's tough to get across to everybody that you have to watch the money. It takes a lot of time raising the money then you have to administer the money properly, then you have to get the right people and of course you must have the product and that's the toughest part."

And, it goes without saying that you have to have patience. "You need the patience of Job if you are in the mining business," says Armstrong. "You have to have deep pockets and the confidence of the investing public that you are going to be successful."

Back in the 1970s, Armstrong started to add oil to his interests. He also incorporated a company called Enex Mines Ltd., and in 1973 started to explore for uranium. It was six years before the Three Mile Island accident and thirteen years before the Chernobyl disaster and countries were searching for uranium for use in nuclear reactors to produce electricity. Uranium deposits turned up in British Columbia and Nova Scotia, but geologists discovered some of the richest properties in northern Saskatchewan's Athabasca Basin. Enex Mines, which would eventually become International Enexco, staked ground throughout Saskatchewan and Alberta searching for uranium. The most promising location they found was at Mann Lake, in the eastern Athabasca Basin.

Cameco and Chevron eventually partnered with Enex in a joint venture for the development of Mann Lake, but then the bottom fell out of the mining industry. Mercury, copper, and nickel prices all fell to a fraction of what they trade for today. Uranium prices dropped from around $26 a pound to less than $6 a pound. In the Athabasca Basin, uranium deposits are usually found anywhere between 500 and 800 metres underground. The huge expense of dropping a shaft down just wasn't economical at those prices. But, while other companies walked away from their claims, Armstrong hung on to Mann Lake.

Over the years Chevron sold out to Cameco and UEM, giving those two companies each a 35% interest and leaving Enex with 30%. First UEM and then Cameco continued to drill, but failed to turn up the high grade they sought. However, early exploration turned up a very large and highly radioactive boulder train indicating the presence of a valuable ore body. "Our geologists were of the opinion that we had uranium there, the question was how do you find it? So we started doing the geophysical work and then we did a limited amount of drilling, but we knew somewhere there was a deposit because of this huge boulder train," says Armstrong.

Armstrong's decision to hold onto Mann Lake would prove very auspicious for the investors in International Enexco. While public outrage against Chernobyl in Russia and Three Mile Island in the U.S. dried up the demand for nuclear reactors through the 1990s, in the last few years this source of energy has come back in vogue. As concern mounts about the environmental impacts of coal, oil, and even hydro power, the clean energy potential of nuclear reactors is garnering them a lot of support. As Paul McKenzie, one of International Enexco's Directors explains there are no carbon emissions from reactors and even some environmental groups are taking up the cause. "They are not all on board and there is some conflict within some of these organizations, but the fact is, if you want to look at a large scale energy project in the near term, nuclear makes a lot of sense. You can build plants that generate a lot of electricity for cities very quickly," he says. "If you look at a lot of clean alternatives like wind and solar they can't reach the vast scale, they can't hit the numbers."

In January 2003 uranium sold for $10.15 a pound. The price peaked at $140 pound in July 2007 as countries experienced a nuclear renaissance. Armstrong says China's demand for nuclear reactors over the next ten or twelve years "will be something in the range of 80 to 100 reactors." The United States is also expected to build a number of new reactors to meet rising power needs, choosing nuclear power to help reduce carbon emissions." Armstrong is confident that the summer 2007 fall-off in the price of uranium and shares of Canadian uranium producers is a temporary price correction. "It's all a question of supply and demand," he says. "There is some question whether or not the demand will continue to exceed the supply, but as far as I am concerned it will." Armstrong predicts that over the next several years the price of uranium will settle at somewhere around $200 a pound. And he notes: "At $25 a pound you can put a uranium mine in the Athabasca Basin into production."

The cost of sinking a shaft into the Athabasca Basin makes production quite expensive, compared to other parts of the world. "It just depends on your deposit, how big it is, what the grades are, how deep it is, what you've got to build, and the cost of mining," says Armstrong. "Each case is different, but the mines already in production in the Athabasca Basin are producing uranium for $6 a pound because they had all the infrastructure already in there." Despite the obstacle, the rewards for mining these deep uranium deposits are immense. Other major uranium producers, like Australia, boast shallow deposits amenable to open pit mining. While much less expensive to operate, there is a huge difference in grades between these shallow deposits and the deeper Athabasca Basin deposits. The grades in open pit uranium mines tend to range from 0.2 to 0.6% compared to the Athabasca Basin where the average grade is 4.2%.

International Enexco's Mann Lake property is just 25 kilometres southwest of Cameco's McArthur River Mine, the largest high-grade uranium deposit in the world. Just 20 kilometres to the northeast of Mann Lake is the Millennium deposit, discovered by Cameco in 2000. The Millennium deposit is expected to prove to be a very rich uranium deposit when production starts in 2008. The Mann Lake property covers the northern extension of a conductor that is spatially associated with the Millennium deposit.

The location, the desirability of the grade, and the results from two drill holes spurred International Enexco to triple its drill program budget from $500,000 in 2006 to $1.5 million in 2007. The two 2006 drill holes tested a 6,000-metre long geophysics UEM conductor at Mann Lake. One hole hit high-grade uranium up to 7.12% eU308 over 0.25 metres just above a second intersection of 5.53% eU308 over 0.4 meters at around 600 metres depth. A second significant UEM conductor was discovered 1,300 metres west of the high-grade discovery zone.

"It's very exciting because of where it is and the size of the grades," says Armstrong. "We are in the vicinity of the highest grade and the most profitable uranium mines on the planet; we are the next door neighbours to these mines. We've hit high-grade, which is very difficult to find in uranium exploration." As Armstrong explains, finding a high grade uranium deposit is like looking for a real needle in a haystack. "So the fact that we hit these grades this early is extremely encouraging. The big question is—is the source big enough to sustain a mine economically?"

The next step is to take that $1.5 million and drill up to nine holes to determine whether there is an actual ore body. "To give you an idea of size, the McArthur River ore body is about the size of a football field— that's all. It's very thick, but it's because the deposit is so high-grade that it's extremely valuable," explains Armstrong. There's also the possibility that the Mann Lake property could yield more than one ore body, but at this stage it's all about possibilities. One of the reasons behind Armstrong's considerable success is his talent at managing risks. While the Mann Lake property could produce Pine Point-like profits, there is always inherent risk. "You don't want to pin your hopes on one thing when you run a company," says Armstrong. "We consider that the uranium side is the up side whereas our bread and butter business really is the copper and silver business in Nevada."

The bread and butter business he speaks of is Contact, a pre-production copper and silver project in northeast Nevada. As with the Mann Lake property, Armstrong staked the Nevada mine back in the 1970s through a mostly oil and gas company that he ran called Coralta Resources. When Armstrong later sold control of Coralta in the 1990s he arranged to buy the Nevada property and put it into International Enexco. Armstrong continued to pay the taxes on the property but essentially mothballed it, waiting for metal prices to improve. Armstrong joint-ventured the property first to Phelps Dodge and later to a company called Golden Phoenix, but after both dropped their options Armstrong staked the entire district. As instinct and luck would have it, copper prices started to rise.

International Enexco is now the sole owner of the Contact copper and silver project and prospects look excellent. The tiny town of Contact is close to Interstate 93 and has links to smelters in Salt Lake City, Utah to the east, and Reno, Nevada to the southwest. Tests indicate over 400 million pounds of copper and 3.7 million ounces of silver from both indicated and inferred categories. As well, 80 percent of the ore body is contained within 300 metres of surface.

The high costs of drilling prompted the company to spend about $500,000 (USD) to buy its own drill rig as well as a new water truck and a 40-foot on-site trailer. The infrastructure makes the company totally self-sufficient. As Art Small, project manager explained, "having our own drill rig and crew makes good sense for us in Nevada. Not only will it allow us to accelerate our current drill programs and make better use of our resources, but in the future it will allow us to target new areas of interest with significantly greater flexibility and savings."

International Enexco raised about $20 million on the stock exchange in three different financings in just over a year and the Contact project is now considered in the advanced stage of exploration. The first of three phases of infill and expansion drilling have been completed at Contact, upgrading the property to the Measured and Indicated resource category. The next step is the pre-feasibility study which will evaluate the viability of various mining scenarios such as open pit mining and heap leach processing versus underground mining and milling. Enviroscientists of Nevada is under contract to undertake and conduct environmental baseline studies and permitting.

Armstrong is confident that the highly qualified team and positive early results will propel International Enexco to be one of Nevada's next copper and precious metal producers. It's a lengthy process, but Armstrong is nothing if not patient. "We have an accessible area that's just off the highway. We have power and water," he says. "Of course we have to be environmentally friendly and those environmental reports generally take eighteen months, and then another one to two years to go through the permitting process. It generally takes another year to bring it into production. So that's our main thrust—to bring that into production at somewhere around a 1000 tonnes a day, and based on prices of $1.25 to $1.50 a pound. It's a very profitable situation."

Ask Armstrong what International Enexco is best known for and he'll tell you it's "tenacity and longevity. You have to be prepared that if there is a failure in one property you are moving on to the next situation. That's what your shareholders are looking for," he says. "In the mining business every 1,000 mineral occurrences will usually result in only one mine. You can see that the odds are terrible. And the other thing is, it used to take about ten years from the date of a discovery to the date of production. That timeline has shortened, but it's still about six years."

One of the ways Armstrong lessens the risk is by hiring the best people. And, it's a testament to the firm's solid reputation that it attracts the best and brightest. "It's all about the people," says Armstrong. "It's knowing the right people, hiring the right people, and letting them do their jobs. I'm better at picking people then I am at running the show myself." People like William W. Willoughby, Art Small, and Mel Klohn all have exceptional qualifications and a belief that, while small, International Enexco is a major player in the industry.

Willoughby, Manager of Engineering and Mine Development at International Enexco caps off his PhD in mining engineering, and MSc and BSc in geological engineering, with 25 years experience in the business, including 17 years with Teck Cominco as a senior mine engineer. Small, the Chief Project Manager, holds a Geology degree from Colorado State University and has worked extensively with both Exxon Minerals Co. and Teck Cominco. Over the years he has worked in drill project supervision, geophysical programs, surveying, claim staking, and property evaluation in copper, gold, and uranium in the United States, Mexico, Chile, and Argentina. Rounding out the management team are Klohn, an Enexco Director, a licensed Geologist in Washington State and a former Exxon employee; Paul McKenzie, Director, and Dan Frederiksen,

Director and Chief Financial Officer. International Enexco also has three crews, for a total of nine full-time drillers, who work in Nevada.

Armstrong says that his corporate strategy in this highly volatile industry is simple, it's all about credibility. "Credibility is very important, because if you don't have credibility then you are not going to get financial institutions to work with you," he says. The idea, adds Armstrong, is to operate small, keep the overhead down, and pour everything into the ground. The company lists on the TSE-V under the symbol IEC, and is well financed at $17.5 million, with an exploration budget of around $3.5 million for both properties. Add in "luck," "great persistence," and "tenacity," and you've pretty much got Armstrong's recipe for corporate success. "It's a very frustrating business, because you may wind up with some mineralization and you raise the money and you drill the property and you don't end up with an economic model, so you have to walk away and go on to the next property and so on. It's hazardous, but that's what makes it challenging and fun."

It doesn't hurt, of course, to have Armstrong's nerves of steel and the instincts of a high roller. George Cross is a resource consultant at Canaccord in Vancouver. For many years he ran the highly regarded George Cross newsletter that his father ran before him. He first met Armstrong in 1964 during the Pine Point adventure and describes him as smooth. "If you are ever going to have a proxy battle," Cross says, "he's the lawyer you'd want. He's so smooth, he's wonderful. Of course he went to school with the Chief Justices and half of the judges in B.C., although I guess a lot are now retired." Cross says Armstrong has also got the personality necessary to survive in his fast-paced business. "You've got to be hard nosed, negotiate hard, play hard, and keep doing it, failure after failure," says Cross. "The rewards are enormous, but you've got to get lucky. And, it's who you know."

Armstrong, of course, seems to know everybody. In the 1970's, Armstrong and his business partner Ray McLean crossed swords with the legendary Howard Hughes. A couple of high rollers, they decided, just like back in the days before Pyramid Mines, that it would be prudent to get on the other side of the table. "I'm not much of a gambler any longer," says Armstrong. "But we used to go to Vegas and we were pretty high flyers on the gambling." As Armstrong explains, at the time Howard Hughes already owned four hotels in Las Vegas when the Landmark Hotel came up for sale. Hughes bid $17 million for the hotel which was a revolving tower with a casino on the top. The Nevada Gaming Commission felt that he already controlled a little too much of the city and let it be known that if anyone else offered $17 million they'd get it.

It sounded like a great bet to Armstrong and McLean. "Ray and I went down to the U.S. and we raised the financing. The Teamsters had an $11 million mortgage and I got them to agree to jump that to $13 million and then I arranged financing for $2 million through another company for all the equipment. We were putting up the balance," he says. Armstrong says everything was set to go until they realized they didn't know the first thing about running a casino. "We weren't experienced and we didn't know the mechanism of running it. You have to have tremendous surveillance equipment and so on, and we thought if we went in we might have a difficult time for the first few months, so much so that we might have been out of business in a matter of months. So at the very end we thought we better not take this gamble and we dropped it."

While that particular deal didn't go through, it was a great introduction and education into the business side of casinos and when three casinos—two in Burnaby and one in New Westminster went up for sale in 1992, Armstrong drew up all the legal documents for McLean and took an interest in the business. McLean set up an income trust called Gateway Casinos Income Trust and added four casinos in the Okanagan—Vernon, Penticton, Kamloops, and Kelowna and then the Palace Casino in the Edmonton Mall came up for sale. It eventually evolved into a billion dollar business with over 2,000 employees. McLean, who was Chair and Chief Executive Officer of Gateway Casinos, retired from that position in 2006 and the following year put the casinos up for sale. He first met Armstrong in the heady 1960s. Ask him the secrets to Armstrong's success and he says "perseverance."

"He's got a good head for business and certainly is willing to accept a challenge. His is very much an entrepreneurial spirit," says McLean. McLean tells the story that in 1967, a few years after meeting Armstrong, he invested in a hotel in the Cayman Islands. "I urged Arnold to buy a piece of it and he did that without seeing it," says McLean. "I don't know if you would call it recklessness, but certainly it proved a good investment. I would say he's got a good feel for business."

McLean argues the resource industry and the gambling industry are both risky ventures and to survive in them you need to be almost foolhardy. "A good measured instinct is necessary, but as long as you can assess the risks you then must decide whether or not it's something that you can live with if you fail. I think Arnold has a fair amount of acumen in that regard," he says. "The end result if you are successful is more than generous. You can reap a much larger return than staying with traditional types of business. I don't know, I guess it's the gamble and the gamble is part of the way some people like to live."

But in the mining business, Armstrong prides himself on balancing risk, as International Enexco is doing with it's combination of an early stage, potentially high return uranium property and a late stage Nevada based copper–silver property. Ask Armstrong what he does for fun, and he laughs. "This is fun," he says. And, don't expect to see him retiring any time soon. "Retire to what?" he asks. "I'm happy doing what I am doing and I work with a strong team that really does 90 percent of the work." With the intention of bringing the 400 million pound copper resource at Contact into production to help offset the cost of continued exploration of the very high-grade uranium mineralization at Mann Lake, International Enexco is well positioned to offer it's shareholders both security and blue-sky potential. And that's a mix Armstrong is happy with.

JNR Resources Inc.

Rick Kusmirski is fond of remembering what happened on the day. He'd been working out of his basement for more than a year, living on his savings. Along with Dave Billard, another Cameco Corporation retread who had agreed to work for JNR Resources on faith, Kusmirski had devoted thousands of hours reviewing historical assessment records, metallogenic models, and the known geology of his old stomping grounds—the Athabasca Basin. What happened on the day was the result of a little bit of luck and a whole lot of experience coming together at the end of a drill bit.

However, long before that day happened, mining promoter Dale Hoffman had some luck of his own, getting Kusmirski and Billard to come work for him. Hoffman was an area player who in 1997 staked more than a million acres of ground. This original land package was largely selected by Dr. Les Beck who was working as a consultant to JNR and is still a technical advisor and member of the board.

Hoffman had spent his life in northern Saskatchewan searching for gold, base metals, and diamonds—actually whatever mineral prospect he thought he could promote—and he'd gotten to know the key 'players' well. "We always had a good rapport," Kusmirski recalls. "At Cameco, we found a number of mineral showings and deposits and Dale used to stake around us as a matter of course," he says. So, apparently Hoffman was following his normal business practice in February 1997 when, with uranium prices forecast to rise after decades of slump, he began to acquire a significant land position in northern Saskatchewan's Athabasca Basin along with Kennecott Canada Exploration Inc. It was before the staking rush occurred in the area, showing Hoffman's visionary style, but interest was building by 1999 and he needed some specialized expertise to maximize his property asset in the Basin. As luck would have it that was about the same time Kusmirski and Billard broke free of secure positions at Cameco and took on the risk and reward of independent exploration. Hoffman offered them both a chance to put what they knew to work for JNR Resources as consultants.

Of all the junior mining companies that were beginning to involve themselves in the rising Athabasca Basin uranium play, JNR Resources was in an enviable position because of Hoffman's land acquisitions. And, by hiring the two ex-Cameco geologists, JNR Resources had snagged some very hard to come by wisdom as well. In actuality, Kusmirski and Billard possessed some of the highest quality technical expertise and practical knowledge about exploration in the Basin. Kusmirski had been Cameco's Exploration Manager for projects there. He worked with Billard who had also spent 12 of his 30 years of mining experience working for Cameco in western Canada and the western U.S.

Hoffman knew the Athabasca Basin hosted a third of the world's uranium mine output and the largest high-grade mine on earth, McArthur River. McArthur River has proven and probable reserves of 0.8 million tonnes at an average ore grade of 25% eU308 and if anyone could find an unconformity deposit that would pay off Hoffman reasoned, it would be Kusmirski and Billard.

Their first focus was a highly prospective piece of ground known as the Moore Lake property. Located only 40 kilometres south of the McArthur River mine and 45 kilometres northeast of the Key Lake mine and mill complex, the property was a prime candidate for exploration. Since the initial staking in the fall of 1999, JNR and its partners have acquired 12 claims covering 35,705 hectares to form the Moore Lake project. After a good deal of study, the geologists recommended targets and JNR Resources announced its first drill program—three holes.

To Rick Kusmirski the size of the spring 2000 drill program at Moore Lake was almost comical. "My last task with Cameco was managing their uranium exploration activities in the Basin and we drilled 100 holes a year. If we had a hit in one or two of them we were pretty excited. With JNR we drilled only three and when the third hole hit it just floored me," says Kusmirski. That third drill hole intersected significant uranium mineralization: 9.2 metres of 0.422% eU308. The mineralized interval was dubbed the Maverick zone and represented one of the most significant discoveries to be made in the Basin over the previous several years.

When Kusmirski called Hoffman in Vancouver to pass on the news, he recalls hearing the blare of car horns. "I said Dale you won't believe it but we hit several metres of mineralization. Then all I can hear are these horns. I asked Dale where he was. Apparently when I told him he'd stopped dead in the middle of crossing Granville Street and nearly got run over."

That was a day Kusmirski will never forget.

Things changed for Rick Kusmirski in January 2001, when Dale Hoffman died. "A few of the larger shareholders asked me to look after

JNR for a year." Although it wasn't his forte being a field geologist, Kusmirski agreed, in part because he knew the years of management experience he gained at Cameco would serve him well. "There was the publicly-traded company end of things and all the shareholder issues to deal with," he says, "but quite frankly that was no different than dealing with partners and the populace when you begin to develop an area." Kusmirski dove in and went to work at a time when the spot price of uranium was floating between $7 and $8 per pound.

With Kennecott Canada, JNR continued to explore. Follow-up drilling in 2002 had confirmed the presence of significant mineralization on the Moore Lake Property. Hole ML-25 intersected 0.62% U308 over 9.1 metres, including a high-grade interval of 12% U308 over 0.4 metres. It was conclusive evidence there was high-grade mineralization on the property. Kennecott, says Kusmirski, "were great partners and they were very fair to us, however they were also looking for world-class Rio Tinto type deposits and they did not believe they could find another McArthur River where we were." By December 2002, Kennecott and JNR had "cut the strings" which left JNR Resources on the hunt for financing. That period of uncertainty displayed Kusmirski and Billard's character more plainly than any other circumstances might. "The junior sector was not a place to be in at that time for a number of reasons. Commodity prices were all low and it was just not an environment conducive to investing. I can remember going to Vancouver to try to raise money and when I got to the airport there were all these people walking around with white masks." Kusmirski remembers the SARS crisis, and coming close to securing financing the day war was declared in Iraq. "There were people reaching for their cheque books and then the war started. Then I was in Toronto and the headline in the paper read Air Canada declares bankruptcy. We couldn't give the shares away at five cents," in spite of holding such a promising piece of real estate in the Athabasca Basin, he recalls.

"We didn't have any cash flow from the middle of 2002 until Thanksgiving 2003. For sixteen months JNR had two employees, myself and Dave. We worked fulltime but JNR couldn't afford to pay us. We were the kind of employees every company would love to have," he says with a chuckle, very easy on the burn rate. "Not only were we not getting paid, the company was about a half million dollars in debt and most of it was owed to Revenue Canada." Kusmirski recalls the struggle of dealing with the tax collector in a marketplace that was soft for financing. "Then we got a call from Lukas Lundin, the Chairman of International Uranium Corporation, (IUC) now Denison Mines. I went to Vancouver and we did a handshake deal, and it's been onwards and upwards ever since. They are a great group of people to work with."

With IUC/Denison Mines as a joint venture partner, JNR Resources was able to promptly initiate a series of aggressive exploration programs on the Maverick zone and numerous other historic and newly identified regional targets on the Moore Lake property. In fact over the past three years the companies have drilled some 30,000 metres a year. Ultimately, the Maverick zone was identified as a major structural corridor, a minimum 6.5 kilometres long. Associated with it is an intense hydrothermal alteration system highly enriched in trace element geochemistry, most notably boron, nickel, and uranium. High-grade uranium mineralization at the Maverick 'Main' zone has now been intersected over a minimum strike length of 350 metres. Furthermore, step-out drilling along the Maverick corridor resulted in the discovery of two new zones of unconformity-style uranium mineralization; the '527' and '525' zones. Together with the Maverick 'Main' zone, mineralization has now been identified over a 1.7 kilometre length of the Maverick structural corridor, over 50 percent of which has yet to be drill tested. "The presence of multiple mineralized zones along a major structural corridor is common to uranium deposits in the Athabasca Basin", says Kusmirski.

The most recently reported results are from the 2006 summer/fall program. Infill drilling on the Maverick 'Main' zone intersected uranium mineralization in six holes with high-grade intersections in three of them. ML-140 returned 3.20% U_3O_8 over 6.5 metres including a 3.5 metre intercept of 5.25% U_3O_8, 2.1% nickel, and 0.65% cobalt. ML-139 returned 1.23% U_3O_8 over 8.5 metres, including a 1.5 metre intercept of 4.20% U_3O_8. The mineralization in both of these holes occurs at the unconformity and in the basal sandstone. Also at the Maverick 'Main' zone, ML-133 intersected two zones of uranium mineralization. A high-grade zone at the unconformity returned 2.72% U_3O_8, 2.30% nickel, and 0.905% cobalt over 5.0 metres, including a 2.0 metre intercept of 4.25% U_3O_8. ML-133 also intersected mineralization in the basement associated with clay-altered graphitic pelites, returning 0.611% U_3O_8 over 3.5 metres. Uranium mineralization was intersected in all three holes that tested the '527' area, with the best result obtained from ML-136 returning 0.50% U_3O_8 over 7.0 metres.

Some of the better results to date from the 'Main' zone include; 4.03% eU_3O_8 over 10.0 metres, including 19.96% eU_3O_8 over 1.4 metres in ML-

61; 5.14% U3O8 over 6.2 metres in ML-55 and 4.02% U3O8 over 4.7 metres in ML-48.

JNR and Denison have also recently being actively exploring a second major zone on the Moore Lake property; the Nutana-Venice structural corridor. This impressive 10 kilometre long, 500 metre wide conductive corridor wraps around the northern boundary of the granitic body that forms the northern contact of the Maverick zone. This corridor encompasses the Nutana, West-Venice, and Venice grids, and is interpreted to be an extension of the conductive system associated with mineralization in the Maverick zone. "The geochemical and geological signatures obtained from holes that tested this corridor are highly prospective, with trace element enrichment, illitic clay geochemistry, and significant graphitic intercepts occurring," says Kusmirski. "These features are consistently associated with major mineralized systems."

Significant results were also obtained from several holes that tested the minimum 4.0 kilometre long conductive zone on the Avalon grid. Of particular interest was ML-850, which represented the first-pass drilling of a 1.0 kilometre segment of the zone. It intersected a broad 25 metre zone of highly anomalous radioactivity located well beneath the unconformity. Extensive structural disruption and/or anomalous geochemistry were intersected in the majority of holes that tested other target areas.

Also of note are the relatively shallow unconformity depths on the Moore Lake property. The unconformity along the Maverick structural corridor varies from 270 to 285 metres below surface. In the Avalon area it is on the order of 320 to 330 metres. The shallowest unconformity depths on the property, around 200 metres, occur on the Rarotonga grid.

A full evaluation of all of these zones is underway, and follow up drilling is planned on a number of highly prospective targets. The 2007 winter exploration program consisted of a planned minimum of 10,000 metres of diamond drilling along with 110 kilometres of linecutting and ground geophysics. A property-wide airborne resistivity and magnetic survey was also in the works. The joint venture budgeted $5 million for exploration in 2007.

Theses results have made the years of struggling for funding well worth it. "We are quite pleased with how JNR Resources has progressed and think that most of our shareholders are as well." Kusmirski says. "It's one of the reasons Dave and I stuck it out in 2002 and 2003. We knew there was something of significance at Moore Lake. There was, in our minds, no doubt about it. We just didn't know how big it was, and quite frankly we still don't. We were confident that eventually we could bring somebody on stream and it turned out to be Denison Mines." Denison has now earned a 75% interest in the Moore Lake property.

"I think that we're still in the position of being quite confident that we can do a lot better based on what we are seeing on our other properties. Now that Denison has a technical group working some of the joint venture grounds, we are able to work some of our 100% owned properties and we like what we're seeing. We feel that we have a very good opportunity to continue to move our share price forward, to our shareholders' advantage."

As well as the Moore Lake property, JNR is actively exploring elsewhere in the Basin. "Our properties are located in high priority areas around the margins of the basin and that's by design." Because the targets being tested are shallower, JNR has been able to carry out drilling programs cost efficiently. "Finding uranium deposits is very drilling intensive, in part because you are drilling blind and your targets are not very large"; and "If you find something of consequence it's nice to know that you have a better than average chance of actually extracting it. That's a big selling point for JNR Resources. Look at the problems they are having now at Cigar Lake."

JNR also substantially increased its land position in the Way Lake area in 2006 in response to obtaining high-grade uranium values and other encouraging results from a reconnaissance-scale prospecting and mapping program. Owned 100% by JNR, the Way Lake uranium project is located 55 kilometres east of the Key Lake uranium mine. It consists of seventeen contiguous claims totalling 71,795 hectares. Kusmirski says that in 2007, $1 million is being spent on diamond drilling and over 100 kilometres of HLEM and ground magnetic surveys on five separate grids on the Way Lake project.

"Of particular interest is the Hook Lake showing, which consists of a massive pitchblende vein, and several proximal radioactive shears and fractures," he says. Two grab samples collected from the vein returned 40.1% and 48% U308, and a soil sample overlying the vein assayed 27.8% U308. The uranium mineralization is also associated with significant lead (up to 8.8%); rare earth elements; thorium enrichment; and anomalous boron, cobalt, and vanadium values. Preliminary results from a 5500-line kilometre helicopter-borne VTEM survey flown in the fall of 2006 indicated over 50 kilometres of arcuate and structurally displaced conductors in the southwest portion of the property. The Hook Lake showing itself occurs in a magnetic low, adjacent to a magnetic high. "Additional groundwork, including drilling, is underway," says Kusmirski.

A new project, Yurchison Lake was also added in 2006. The Yurchison Lake claims are owned 100% by JNR. They were staked contemporaneous with the acquisition of the additional Way Lake area claims in response to significant uranium and molybdenum results being obtained from both outcrop and float samples.

JNR owns a 100% interest in three additional properties in the Athabasca Basin; Crackingstone, Black Lake and Newnham Lake. All three have seen various levels of exploration over the past couple of years and all three are slated to be covered by a high resolution airborne gradient magnetic survey in the fall of 2007. In the case of Black Lake and Newnham Lake, the results from this survey will be collated with previous airborne and ground geophysical surveys, to plan follow-up drilling campaigns during 2008.

Other Atahabasca Basin projects of particular interest are three Joint Ventures with Denison; Bell Lake, Lazy Edward Bay and Pendleton Lake. JNR holds a 40% interest in the Bell Lake project, where a diamond drilling program planned for the fall of 2007, will follow-up an extensive EM/magnetic conductive trend identified by airborne and ground geophysical surveys. Drilling is also planned for the fall of 2007 on Lazy Edward Bay and Pendleton Lake. JNR holds a 25% interest in each of these projects.

"Currently, we have plans to carry out exploration on eleven projects in Saskatchewan with drilling planned on five of them by the end of 2007."

"We also have the South Fork project located east of the Cypress Hills in southwestern Saskatchewan," Kusmirski says. In May 2007, Uranium Power Corp (UPC) signed an option agreement to earn up to a 65% interest in the 25 mineral claims and one mineral permit that make up this project. Along with other conditions, the terms call for UPC to carry out $1.5 million worth of exploration before the end of January 2009 to earn the first 50% interest in the project. UPC can then increase its interest to 65% by spending an additional $1million on exploration activities by January 25, 2011. The South Fork project, first explored in the late 1970s and early 1980s by Saskatchewan Mining Development Corp (SMDC) comprises just over 50,000 hectares with excellent potential to host roll-front uranium deposits. The geological environment is very similar to the prolific Powder River Basin in Wyoming. In Wyoming, such deposits are commonly amenable to in situ recovery.

"The opportunity to work with UPC affords us the ability to utilize UPC's operational expertise in the exploration and development of in situ recoverable uranium deposits," Kusmirski observes. He says the decision to joint venture was also an issue of JNR's manpower. "We do have a lot of good technical people and we're able to do a fair amount of work. Our Vice –President of Exploration worked in situ recovery operations for Cameco in Casper, Wyoming for four years, so he's quite adept at it but he's better utilized elsewhere."

In addition to the projects in and around the Athabasca Basin, JNR has an option with Altius Minerals to earn a 70 percent interest in their Rocky Brook uranium property. The Rocky Brook property is located in the Deer Lake Basin in western Newfoundland, and features three distinct areas of un-sourced altered and mineralized sandstone boulders with reported values ranging from 1% to more than 10% U308, as well as very high-grade silver contents. A successful reconnaissance-scale drill program in 2005 identified areas of alteration and geochemical enrichment analogous to the mineralized boulders. Significant radiometric anomalies were also identified in the glacial till overburden. In 2006, 65 holes totalling 2,881 metres were drilled. Uranium mineralization was intersected at shallow depths in a number of holes in the Wigwam Brook area. RB-06-117 intersected a grade equivalent of 0.54% U308 over 0.10 metre within a 0.40 metre interval of 0.075% U308, and 4.4 ppm silver. RB-06-127 intersected a 0.5-metre interval of 0.80% U308, 1030 ppm copper and 2.2 ppm silver. Anomalous radioactivity was also encountered in several holes drilled immediately northwest of the Birchy Hill Brook boulder anomaly.

JNR is also looking to expand its portfolio of properties. "We're actively looking now. JNR's future will be very much knowledge driven, according to Kusmirski. "Of course, having Ron Hochstein as a director is an obvious plus." Hochstein is the President and COO of Denison Mines Corp. and had been the President and CEO of International Uranium Corporation for nearly six years prior to that.

"The experience we have on our technical team far exceeds just knowledge. It moves into the wisdom end of the technical spectrum. For example, Dr. Les Beck brings with him some world-renowned expertise on Saskatchewan uranium deposits." Dr. Beck was formerly the Executive Director of the Geology & Mines Division for Saskatchewan Energy & Mines.

Another strong member of the JNR management team is Tracy Hurley who has more than 25 years of experience in the mining and exploration industry, including six years as a mining analyst for a Vancouver based brokerage house. Her technical expertise in a supervisory capacity spanned grassroots exploration to multi-million dollar mining and development operations.

"And when Ken Wasyliuk signed on with us a couple of years ago it was a big coup." Wasyliuk, JNR's Chief Geochemist, has over 20 years of industry experience including 16 years with Cameco. At Cameco, Wasyliuk became an expert on geochemical and clay alteration patterns associated with uranium deposits in the Athabasca Basin. "Ken is one of the few people in North America who can carry out Portable Infrared Mineral Analyzer (PIMA) work. Not only does he do that for us but we subcontract PIMA analysis to several other juniors and generate cash flow as a result, which helps defray office and administration costs."

Kusmirski considers the appointment of Dr. Irvine R. Annesley as Director of Exploration at JNR as another coup for the company. As a geologist, Dr. Annesley has 30 years of field experience, the past 19 as a senior research geologist with the Saskatchewan Research Council (SRC). During his time at the SRC, he worked closely with the exploration and mining industry. His research focused on uranium deposits in the Athabasca Basin using conventional geological methods such as drill core logging, petrography, whole-rock geochemistry and uranium-lead geochronology. His research also involved innovative techniques including advanced structural lineament analysis, applied synchrotron research, GOCAD structural and geochemical modeling, numerical modeling, lead isotopes and 3-D visualization.

"Irv's appointment further enhances the expertise of JNR's technical team and it demonstrates our ongoing uranium exploration commitment. His addition also allows us the latitude to expand our activities and advance our projects. For instance, Irv always described Maverick as very fertile property," Kusmirski says. "He believes that with some knowledge-based targeting, you just can't drill dead holes. There is always something kicking and for us that means there is just so much more to test." He says that goes for the entire property. "I think there are more than a dozen conductive systems on the Moore Lake property, and we've really only focused on three or four of them. And, we are still carrying out ground geophysics and identifying untested targets."

Kusmirski's first love, geology, has had to take a backseat to the onerous job of running one of the most active junior exploration companies in the

Athabasca Basin. Oddly enough, "a perceived lack of management activities was one of the reasons he thought that he would enjoy working with a junior company once he left Cameco," he says, laughing. "I sort of went from the frying pan into the fire, all right. Every year there is clearly more and more paperwork, legal documents, etc, associated with running a junior company" he says. "So it really draws me away from the geological end of things. But if that is what you have to do, then that's what you do. I have to admit I miss the geological aspect of exploration and that's why I am very pleased to have these good technical people here. I always defined exploration as brief moments of excitement surrounded by endless periods of tedium and frustration. People say why do you do it? It's because of that brief moment like the day back in April 2000. You just feel your heart pounding, and there is no feeling like it."

As far as the tribulations of a hot exploration environment hampering the activities that JNR so methodically plans, Kusmirski says the company has no difficulties. He suspects that is due, at least in part, to the cumulative relationship pool that the directors and key management have been able to create over their many years working in Saskatchewan exploration. "It hasn't been hard to find the rigs. We've been active in this province since the mid-1980's so when it comes to knowing the contractors, the linecutters, the geophysical contractors, and even the drillers—we do. I mean, we supported them when times were lean and now that times are good they've opted to stick with us. And we know each other well, so there are not a whole lot of hassles when it comes to determining costs and interpreting contracts and things like that."

As far as dealing with the many stakeholders involved in creating a mining project, Kusmirski says dealings with First Nations peoples have been very good. "There have been no problems. We have a duty to consult with the aboriginal community, and we do. Sometimes it slows the process down a little bit, but quite frankly the permitting process is not overly cumbersome. Compared to some other jurisdictions that I've worked in, permitting exploration in the Athabasca Basin is not a problem. We employ a large number of individuals from the North, both directly and indirectly, and they do very good work" he adds proudly.

And reflecting on the future of uranium, "There is currently a bit of a fall back in the price after going virtually straight up" he says. "After three years of appreciation, a down tick isn't unexpected. Even with a major stock market correction, the uranium price has remained relatively firm and I feel that it will stay this way. We are still in an era with a significant gap between production and demand so, although some experts are saying that the uranium and commodity bubble are going to burst, I don't agree." He points to China as the strongest proof for his argument that uranium will continue to be the darling of commodities long into the future. "China used to subsidize exports of metals such as tungsten, antimony, etc., but they are going to be net importers now. Just consider the number of uranium reactors they are planning to build. Starting in 2010, China is talking about bringing one on stream every couple of years.

"Then there is India and other growth countries. The demand is just going to continue to move upward and the supply is going to take a lot longer to get on stream than anyone expects. Cameco has a uranium deposit called Millennium that is located some 30 to 35 kilometres to the west of the Maverick zone. It has been estimated they have about 50 million pounds in it and they are going to the feasibility stage, but it could be ten years before they can actually extract any ore. Based on all those factors, I think that you are going to see the price for uranium continue to stay firm and possibly move upwards even further."

On the subject of JNR Resources taking a mine to production, the geologist shakes his head. "Our exit strategy today is the same as a lot of juniors. We want to find something of consequence and then get taken out by a larger company. You see the greatest appreciation in share price during the discovery stage. Once you've made the discovery and said 'okay we'll go to feasibility' then basically you're looking at a flat lining of your share price. At that point you are talking several years until you get to the mining stage while you're expending funds and really getting little in return. If we can prove up as little as 30 million pounds, rest assured a mining mogul will take us out."

Mawson Resources Ltd.

S it down to talk about the business of uranium exploration with the men behind Mawson Resources Ltd and you quickly realize something unique is going on. In the first place, these fellows honestly like each other. They embrace the rough edges of their individual characters and in doing so have found a comfortable compatibility that's working wonders as a team dynamic. The fact that the trio of geologists heading up Mawson are all Australians may have something to do with it. As entrepreneurs, they're hived from the same risk-taking mold, and share a brassy, in-your-face kind of honesty.

The Mawson Resources office twelve storeys above West Georgia Street in Vancouver, is not in the absolute centre of the financial district. It's a few blocks away, set a little apart from the central hurly-burly just as the men of Mawson are a little different from the average junior exploration company managers. Hudson and Henstridge share an office with a wall of glass that opens to a view of Burrard Inlet, but the views from their desks are much less picturesque. In the old banker's desk fashion, the two men have their workspaces butted together. The physical layout of their office promotes intellectual intimacy and the trading of knowledge between Hudson and Henstridge. In and of itself, that is unique.

These men aren't typical explorers or miners. They'd be better described as re-explorers. As risk managers they excel at finding properties with historically proven value, re-defining the asset, and nursing it to pre-feasibility.

The management of this dynamic uranium exploration enterprise has found a way to blend cultural differences and on-the-ground knowledge of foreign uranium territories with the "Mawson way of doing things" in order to build an international project portfolio. Mawson has become a legitimate global player in uranium, with projects in Sweden, Finland, and Spain, and the management is as comfortable talking about those countries as their homeland. Through the application of technical knowledge, the acquisition of savvy employees with experience in those foreign lands, and a liberal dose of dedicated labour by everyone concerned in the Vancouver head office, Mawson Resources has become an innovator. Mawson is notably closer to the feasibility stage on projects, particularly one in Spain, than most of their other junior exploration counterparts on the TSX.

At the top of the management pyramid and founders of Mawson are the Aussie triad of Michael Hudson, President and CEO & Director; Mark Saxon, Director & Vice President of Exploration; and David Henstridge, Director. Rounding out the Board of four and also a founder of Mawson and long time associate of Henstridge is Canadian, Nick DeMare, a chartered accountant who provides accounting, management and securities regulatory compliance advice as well as a firm knowledge and understanding of the North American equity markets. He graduated from the University of British Columbia in 1977 with a Bachelor of Commerce.

Hudson and Saxon are both in their late-thirties and graduated from the University of Melbourne geology program a year apart, both with B. Sc. (First Class Honours) in Geology. Henstridge is a veteran of the Vancouver equity market and gold plays in South America. He is a Fellow of the Australian Institute of Mining and Metallurgy, and a member of both the Australian Institute of Geoscientists and the Geological Society of Australia. He graduated with a honours degree in geology as well, but from Adelaide University, and spent his early career years doing field exploration work in Australia.

Henstridge had built an experience bank account of 15 years in mineral exploration in Australia, Papua New Guinea, Europe, North and South America, Fiji, and China by the late 1980s. When the 1987 stock market crash brought about a demise of the junior exploration industry in Australia, things changed dramatically for the Melbourne geologist. Sourcing consulting work or raising money Down Under became next to impossible. Employment conditions were so bad at the time that the business section of national Australian newspapers joked about geologists and the mining sector as a bygone industry—

"What's the difference between a geologist and a four-slice pizza? The pizza can feed a family."

Henstridge suffered through those dismal conditions for several years before he got a crack at coming to Canada near the end of 1992 with a job for Kookaburra Resources Ltd. Kookaburra, a Vancouver subsidiary of Xenolith Gold, was run by Graeme Robinson. He helmed Peruvian Gold Limited from 1993 to 2001. It was in 2002 when Henstridge teamed with Harvey Lim and Nick DeMare —chartered accountants in Vancouver— to take a capital pool company they had created in 2000 called Tumi

Resources and turn it into an exploration company on the hunt for silver in Mexico.

Australians, no matter where they settle temporarily for work around the world, always seem to go home at some point and that is what Henstridge did after nine years. He returned to Melbourne, worked in Mexico, and commuted to the Vancouver office digitally.

For Hudson and Saxon, geographical transitions were a little different. After graduating from university, the two men worked for Pasminco Australia Ltd. Michael Hudson, an affable and observant 38-year-old geologist, developed his exploration to pre-feasibility project management experience in Pakistan, Australia, Mexico, Argentina, and Peru, and for a time, he headed a team on the hunt for mineralization in the arctic areas of Sweden in alliance with BHP Billiton. His discoveries include the Portia gold deposit in the Olary district of South Australia. Mark Saxon, a year younger than Hudson, worked for Pasminco in Australia and South America, shoulder-to-shoulder with his friend.

"We ended up in Peru together in the late 1990s," recalls Hudson. "We were sitting on top of a very big and cold hill, when we decided we wanted to be entrepreneurs. It was the worst time to start on our own but we thought it may be a good time because the bad conditions meant we had a lot of motivation to succeed," he says, smiling. "I think it was a good way to learn, with a few hard knocks." The international exploration company the two men were working for "went belly up," says Hudson, which gave them the opportunity to purchase all the company's data throughout Peru. "We bought it thinking we'd try to commercialize it if you will, and brought the data back to the east coast of Australia.

"Everybody told us the Peruvian project we had was really a Canadian deal and the only person they could think of with experience was a fellow named David Henstridge because he used to have a company called Peruvian Gold." Michael and Mark undertook a global search for Henstridge, only to discover "he was around the corner from us in Melbourne and had just moved back from Vancouver. Over the course of a few years, the young men got to know the older geologist. "We'd found a mentor, we had many balls in the air and we decided that when the market started bouncing we'd come together and he accepted us into the fold. That made for a much easier entry into the Vancouver market."

Henstridge adds, "We must say at this point of time that none of what we are talking about would have been possible without the 100% support of our wives, Sandra Henstridge, Paula Saxon and Debi Hudson.

Geologist's wives are always left to nuture the family and look after home base while their geologist husbands are roaming the world."

In late 2003 the decision was made," remembers Henstridge. Mawson Resources commenced operations on March 10, 2004. On October 28, 2004 the company completed its initial public offering, led by Ian Gordon, who at the time was working at Canaccord and on the back of an Independent NI43-101 techincal report by Andrejs Panteleyev Ph.D, P. Eng. On October 29, 2004 Mawson commenced trading of its common shares on the TSX Venture Exchange under the symbol MAW. At the end of March 2005 the company also listed its common shares for trading on the Frankfurt Stock Exchange under the trading symbol MRY. "With the structure we set up we've successfully built upon IPO stages from a company with a market cap of $5 million to one of $70–$80 million in the space of three years."

"It's a very complicated global company," says Hudson. "We live in Australia, we're listed in Canada and Frankfurt, and our assets are in three European countries, so we employ Canadians, Swedes, Finns, Australians, Peruvians, and Spaniards. Even our German marketing arm is well lead by Pascal Geraths, who is based in Austria."

Initially, with the help of primary industry specialist, Folke Söderström, who Hudson had worked with in Sweden since 2001, Mawson acquired gold assets in Sweden, but with the sudden global spiking of interest in uranium, the team decided to cautiously change focus. "We were still maintaining our gold focus," says Hudson. We didn't know that uranium ever would go where it has."

The team began assembling a portfolio of Swedish uranium projects before most other companies took an interest there. "At the time, we wondered if we should stake for uranium because the real upswing in the uranium cycle wasn't underway at the time. But we decided the properties we were interested in were strategic resources. We thought, 'Let's stake them and worry about them later,' and we did a very good job. We probably staked three-fifths of what was left in Sweden. There were a few leases already gone, but we certainly got some of the most significant resources," Henstridge says proudly.

Sweden currently generates 50 percent of its national electrical requirement from ten nuclear power plants. With the European Union strongly supporting domestic production of U308, and Sweden potentially sitting on large uranium resources it was amazing there was only one other company at the time that Mawson had to compete against.

Sweden's energy supply is 47 percent nuclear, 44 percent hydroelectric, and 9 percent biofuel. Sweden ranks among the top countries in the world for reliance on nuclear power, as well as for power plant efficiency. The current government recently overturned a 27 year anti-nuclear power policy, and has approved the expansion of some nuclear plants. In Sweden, the same legislation is applied to all minerals, including uranium, during the exploration phase. To gain approval for uranium mining, a company must apply to the national government for a ruling. There is no ban on uranium mining in Sweden, and the current government has stated it will review all uranium mining projects in light of the relevant legislation and environmental standards. However, the municipal government where the specific project is located retains a right of veto for uranium mining projects. Sweden's nuclear power plants require 1,500 tonnes of uranium fuel each year, all of which is currently imported.

Just across the border in Finland the world's largest nuclear reactor and Europe's first to be built in 15 years, was under construction at Olkiluoto 3—couple that with the fact that Sweden leads the world in nuclear efficiency research and nuclear waste disposal, and Sweden seems an obvious choice for exploration and mine development. Nonetheless, Mawson was ahead of the curve there. It wasn't luck in Henstridge's opinion. "It is absolutely because the guys around here are astute. They understand that you have to get the right advice in these new countries you go into. If you dig long enough and hard enough, you'll find the right person or the right way to do things. That dedication and professionalism is what makes our luck," he says.

To acquire their properties, the Mawson team carefully studied the historic resource data and Hudson says Mawson was able to cherry-pick what they felt were the largest and highest grade projects at the time. "The data was literally in cardboard boxes that hadn't been looked at since Sweden spent $46 million exploring for uranium in the 1970s," he says. Although none of the historical resource estimates were NI 43-101 compliant, Hudson and Henstridge were excited about the potential. Henstridge was one of the team who discovered the Ngalia basin uranium projects and occurrences in Australia and the Swedish numbers told them these were uranium opportunities in waiting.

They settled on 15 projects, with the key projects being Tåsjö, Kläppibäcken, Duobblon, and Flistjärn. Relying on the experienced field hands of Andrew Bradley, Dylan Jeffriess, Geoff Reed, Rasmus Blomqvist and Lars Dahlenborg, exploration started in earnest in early 2006.

The Tåsjö exploration target is estimated between 104 to 116 million pounds of U308 and 165,000 to 180,000 tonnes of rare earth elements (REE); Duobblon had a Canadian Institute of Mining and Petroleum (CIM)-compliant inferred resource of 11.56 million pounds U308 and is the largest drilled out resource in Sweden; Kläppibäcken's historic resource came in at 2 million pounds; and, Flistjärn showed a series of high grade uranium mineralized structures sampled within an area of 450 metres by 600 metres that assayed from 0.01% to 19.1% U3O8 and averaged 1.7% U3O8. Where channel samples were taken, the mineralized structures were sampled across widths which varied between 0.2 and 0.5 metres.

Tåsjö, located in the Jämtland and Västerbotten counties, is 200 kilometres north of Östersund in northern Sweden. The target estimate there is based upon 100 previous drill holes. A three to ten metre thick, uranium-mineralized, phosphatic calcareous shale exists over a vast area at the Tåsjö project. "Tåsjö is still in the early stages, but it's the big one. An exploration target of 75 to 150 million tonnes has been estimated at a grade between 0.03% and 0.07% U308," Hudson says.

The magnitude of the exploration target was confirmed in a 2006 independent NI 43-101 technical report by Andrew Browne of GeoSynthesis Pty Ltd.—the current qualified person at the Jabiluka uranium project in Australia—after a review of Swedish Geological Survey documentation, a field visit, and check analysis of core samples.

Uranium at Kläppibäcken is hosted by brecciated fluorite-rich granite at surface. The breccia is 50 to70 metres thick. The CIM-compliant indicated resource there is 2 million lbs at 0.1% U308. Drill intersections include 53.1 metres for 0.16% U308 from 30.2 metres and 42.7 metres for 0.11% U308 from 2.8 metres. "Kläppibäcken has shaped up more," says Hudson. "From a mining and exploration point-of-view, it looks like the most favourable project we've got in Sweden. We understand a lot more about it."

Expanding its Scandinavian portfolio, Mawson also acquired a 100% ownership of claim application Nuottijärvi 1. Nuottijärvi is one of Finland's largest known uranium deposits and was identified in 1959 from the discovery of a radioactive outcrop.

Mawson used the same "re-exploration" tactics, along with the experience and savvy of Maximo Casañ Daries, to gain a major place in the uranium space in Spain as well. "Casañ had 35 years understanding of the mining law and exploration issues in Spain, and it was through our sleuthing with him that we found the Don Benito near-mine project. Because of Max's understanding, we discovered we were able to stake the property," Hudson says. Casañ together with the experience of PhD mining engineer Arturo Gutierrez del Olmo and Max Casañ Bates, were in Hudson's opinion, instrumental in helping Mawson move closer to pre-feasibility with a project than the company has ever been.

Spain produced uranium as recently as November 2002 and is a significant uranium consumer with eight nuclear reactors providing 23 percent of the national electricity production. The Don Benito Uranium Project is one of two principal historic uranium mining and processing regions in Spain. Applications, or "Permisos de Investigación," cover three historic project areas: La Haba, Corredor De La Guarda, and Las Cruces-Manantial. The La Haba Project Area includes an historic open pit uranium mine. Total mined and in situ historical resources at La Haba total 9.4 million pounds of U308, of which 6.0 million pounds at 0.06% U308 remain in situ. These resources are not NI-43-101 compliant. Hudson says the permitting of the Permisos de Investigación are underway and proceeding as anticipated, which could mean permits will be granted by December 2007.

The Don Benito claims cover a 35 kilometre trend along strike, to the east and west of the mined area. The claims and follow the contact between a granite pluton and a 300 metre wide black shale unit that is host to the uranium mineralization.

"The 100 percent acquisition defined a major step for Mawson Resources, and expanded our uranium assets into a third European country. Don Benito is the first project we'll take through to pre-feasibility and then feasibility," Hudson explains. "If all is viable and permitted, we'll be producing uranium there. It is a high merit project as there are not many uranium projects that have infrastructure and resources in place.. It's got the roads and it's got the power. It had a mill, but that is no longer in place. But it is a near-production scenario. Under the direction of experienced geologist Sergio Tenorio, we are fairly confident we'll have that permitting in place fairly shortly and we're planning to have a scoping study finished in the first quarter of 2008."

Hudson is confident about Mawson's position in the uranium business. "We have real pounds, which is one compelling reason Mawson is a wonderful investment prospect. Another is that we've got the people in place with the right technical background. We're a technically driven company. And finally, a good reason to look at Mawson as a shareholder is we have the money in the bank to undertake extensive exploration. We are well set to move forward, advancing our current projects with drilling and acquiring new ones. By the same token, it's important to remember we're not just looking for uranium. We've found it, and are making it bigger, better defining it, and solving metallurgical questions—those more advanced questions you'd expect an established explorer moving to pre-feasibility would ask.

"Our goal has always been to find a project that's mineable via discovery and/or acquisition. What we're doing now is advancing the resource projects we have, making them bigger and giving them weight. When they get to a certain size, we'll complete economic studies. This industry would sometimes rather have a hot bonanza hole," Hudson states, "than a deposit that will grow. That is because a hot bonanza hole is more easily understood by shareholders. But we're miners. We're serious people and we aren't trying to ramp a stock on a rock chip or one hole result. We want to create assets and underlying value. And we always have to keep in mind where we're going with that too, because Mawson could become quite a boring stock if we just went down the feasibility path." He says advanced exploration at Don Benito is an example. Results at Don Benito indicate there are 35 kilometres and 30 projects along strike. "To keep Mawson exciting to investors we need to show activity. For example there could the grand daddy of Don Benito out there. You always have to have new things coming on stream.

"If you want to be successful you must always be on the hunt, as only 1 in 10,000 projects ever becomes a mine." Henstridge adds. "You may have ideas that never come to fruition but you can never stop testing those ideas. That applies to the explorationist in us, that part of us that always wants to find a new deposit from scratch. That's where we always get our biggest kick from so we're always looking." In another book that highlighted his career—Gold Rush—Henstridge confided it was the mystery of mineral exploration that keeps him active in the industry. "The thrill of finding that next deposit. That's the reason I'm still here today, the reason I have to be here today. You really get that taste in your blood when you are trying to find a deposit in the earth—it's a bit like hunting for buried treasure. There is no doubt about that, once you're an explorer you're always an explorer and you're always trying to find the next deposit."

Henstridge and Hudson also agree that the key to the future potential of Mawson will originate with its staff. "The biggest asset of mining companies today is the people," Henstridge explains. "This industry has exploded in terms of the sheer number of companies, but the number of experienced people is dwindling. The strength of the technical team at Mawson should be noted. The staffing is impeccable and for management to be successful it needs a supporting cast. The employee quality has to go right down to the bottom level, as it does at Mawson, in order to succeed. No one in their own right can make it happen," Henstridge says. "I'd be remiss if I didn't mention my personal assistant Mariana Bermudez, as an example of that. She has been my colleague since the days of Peruvian Gold going back 13 years." Bermudez is corporate secretary for Tumi and Mawson and Henstridge says she looks after all the material that comes in and goes out of Vancouver, a vital task considering the globe-trotting nature of the management's residence and exploration activity.

"Mawson has a very strong upper management team of geologists and accountants who have driven this success story. These are people with a background of actually finding deposits and taking them to a level where other people are willing to take them on. Mawson is not just one competent person surrounded by a lot of young people who don't have the experience," explains Henstridge. "The average age of key people in the Australian mining industry is 55 years old, so there aren't a lot of long-term survivors like me," he says with a chuckle. Then, more seriously he squints across the cluttered expanse of their two desks which have been butted together and, almost as a warning to his younger partner, adds that he thinks there will be even fewer survivors soon.

Hudson agrees. "I think a shakeout is happening too. There has been such a large amount of money circulating and such a large number of junior companies created in the sector that some changes are bound to be necessary. There was certainly over exuberance in the market, and though I think the long term fundamentals for uranium are there, there will be a consolidation of the real assets in the business to form a smaller group of larger companies." Mawson, he believes, is well positioned for the consolidation phase he's predicting for the uranium industry.

"I don't think we'll ever see the price of uranium regress to what it was ten years ago," Henstridge adds, saying uranium has a future because of economic factors that can't be stalled. "Australia had a visit from the head of China not that long ago. He made the off-the-cuff remark that as far as he was concerned China would buy every pound of uranium that Australia could ever produce. He believes that to make his country grow from an energy viewpoint, China has to build a brand new nuclear reactor every year for the next 50 years. Now we can't comment on whether that is possible, or feasible, but there is an underlying truth in what he says.

Nuclear energy is going to have to fill the void between oil supply and the demand of the developing economic super-powers of China and India."

The winners in the uranium space—those who can go from exploration, to asset, to development to mining—will be the companies with a social conscience. Henstridge says Mawson "will always look at projects in light of the social consequences going forward" for that reason. Hudson argues the swing in attitude towards mining may soften because of the changing economic realities of energy supply, but mining development will not get any easier. "Society's reaction to mining has changed to the point where now, as miners, we don't have the mandate to operate. We have to earn the mandate. That sounds like a company line, but it is true. To succeed, you have to integrate the community at the very earliest of stages, involve them in what's happening with full disclosure." Aren't all exploration companies doing that? Hudson's face shows a small grin. "I think there will be a lot of surprises going forward in delays to uranium projects because of the emotion and the politics that surrounds them," he says. Many companies often fail to invest the time and effort into the public education and integration process in their areas of interest. "They'll discover the fact you need the energy from uranium doesn't really matter as much as the emotion and politics," he says.

Hudson admits that the complex nature of Mawson Resources and the information-packed descriptions of activity the company provides to shareholders on the corporate web site can mean some people find it difficult "to get their head around Mawson. It's always a struggle for us and our IR consultants, Nick Nicolaas and Wayne Melvin from Mining Interactive, who have been utterly dedicated to branding the Mawson story in North America since the company's inception. As professional explorationists we understand the need for a portfolio of projects and to spend the majority of our money on the best projects, however we must concurrently explore some earlier stage projects to bring them up through the pyramid. We're not a single project company because the risk is too high that way. We do what we're trained to do. We manage exploration risk by working a portfolio of projects, but that doesn't always help in creating a single vision of the company in the marketplace. In the same way we have developed a number of key joint venture and equity partnerships to share that risk."

In spite of all the hurdles environmentally, financially, and socially, however, the management of Mawson Resources is confident Mawson is well placed to develop its exploration portfolio towards the sustainable production of uranium in the shortest possible time frame. They are living up to their corporate objectives: To build a mining/energy house and to deliver value to shareholders through rapid discovery, acquisition, and development of uranium deposits.

And they are having fun? "Of course," says Henstridge. "We're Australians. We hate boring."

Pan African Mining Corp.

Madagascar was just an island in the Indian Ocean for Irwin Olian until he got a call from Chris Dunston, a senior executive with CARE Madagascar, telling Olian he had to see what was taking place on the island.

It was 2002 and Madagascar had recently emerged from under the heel of socialist dictator Didier Ratsiraka. Though economic reforms had been slowly taking shape since Ratsiraka's initial loss of power in free elections at the beginning of the 1990s, the aftermath of elections in 2001 made it clear that the Malagasy people were keen on embracing freedom. When Dunston called Olian, he brought news that the economy was opening up, a new mining code was being drafted and—in a word—opportunity was knocking on the door of a country where half the population live in poverty.

Dunston was calling on Olian, whom he knew through mutual friends, to help make some of those opportunities concrete for the Malagasy people.

"Hey, you've got to come down here to Madagascar," Olian remembers Dunston saying.

The new freedom was spurring opportunities not only in traditional commodities such as sugar cane, vanilla, spices, rice, and fish, but also in mining and other resource-based industries.

Soon, despite an initial skepticism, Olian found himself on a plane bound for Madagascar, the world's fourth-largest island located 400 kilometres off the coast of Africa across the Strait of Mozambique. The trip was the beginning of what Olian called "the next great big adventure of my life." Shortly after returning to North America, Olian began moving forward with plans for Pan African Mining Corp., a Vancouver-based junior mining company focused on exploration work in African countries Olian believes, "have relatively low political risk and good geological environments."

Now, five years after Olian's initial visit to Madagascar, Pan African is booming. The company employs some 80 people full-time in Madagascar, virtually all of whom are Malagasy—as residents of Madagascar are known. In addition, it has a dozen highly experienced expatriate geologists on staff and has personnel in its Vancouver headquarters, as well as in Botswana, Namibia and Mozambique.

Among the opportunities capturing Pan African's attention are Madagascar's rich uranium resources. "The uranium project is one of our most exciting projects," Olian explains. "It's advancing rapidly, and we're getting excellent results." Madagascar is the only country where Pan African is pursuing uranium, but it has other ventures in Botswana, Mozambique, and Namibia – a good start, Olian says, towards fulfilling the destiny encapsulated in its name.

Pan African is one of a growing number of companies currently exploring what Madagascar has to offer, and with good reason. From the 1930s through the 1950s, while it was a French colony, Madagascar was an important uranium producer providing supply for France's nuclear energy industry. Many of the old French claims have returned to the limelight with the economic freedoms of the post-Ratsiraka era and the rise in uranium prices. Pan African has a first mover advantage over many of its competitors, having acquired mineral licenses in Madagascar shortly after the new government came into power. Included in its extensive land position are over 20 distinct mineral properties. Among them are many of the old French uranium mines as well as other new and untapped sites within the company's licensed area.

Olian says the company has chosen wisely, focusing its uranium exploration efforts on the Tranomaro region in the south of the country.

"There's a very rich body of uranium ore that is in the form of uranothorianite," he says of the Tranomaro area, where Pan African holds licenses to 1,287 square kilometres. Olian feels the area should yield a good number of productive ore bodies which can potentially be joined with a single large regional processing facility. The most productive uranium area encompasses just 24 square kilometres. "We have chosen to develop that deposit first because the grades are so high and it covers such a large mineralized zone," he says.

But the Tranomaro property is not the only one Pan African holds. In aggregate through its operating subsidiary PAM Madagascar Sarl, Pan African has obtained 55 exclusive research permits encompassing 1,161 carrés. A carré is an old French measure of about 6.25 square kilometres, or approximately 625 hectares. In aggregate, the exploration area Pan African has under license in Madagascar covers approximately 7,500 square kilometres of diversified mineral properties which encompass potential gold, coal, precious stones, base metals and industrial commodities and 5,000 square kilometres of uranium properties. Such claims reflect input gathered by Pan African's geological team from both personal visits and superficial examination; information published in the Madagascar mineral inventory, data acquired from the Geological Survey of Madagascar undertaken by the French government agency Bureau de Récherches Geologiques et Minières over a period of four decades, as well as anecdotal evidence obtained from local inhabitants concerning discoveries in their immediate environments.

A lengthy selection and approvals process meant it wasn't until two years after Pan African's formation that an agreement was signed with the government of Madagascar allowing exploration work to begin for uranium. When it did happen, at the height of the Malagasy summer on January 25, 2005, the state stepped in as a joint-venture partner. Under a preliminary agreement with the Office des Mines Nationales et des Industries Stratégiques (OMNIS), a government agency of the Malagasy State charged with oversight and administration of the country's strategic resources (uranium and hydrocarbons among them), Pan African agreed to the establishment of a joint venture for the exploration, development, and exploitation of four prospective uranium properties in Madagascar. In accordance with the terms of the agreement, Pan African organized a new operating subsidiary named PAM Atomique Sarl, of which a fifth is to be owned by the Malagasy State through OMNIS and 80 percent by Pan African. Approval of Pan African's formal exploration program and budget was specified in the agreement as a condition that had to be met prior to its moving forward with the joint venture's uranium exploration and development activities through PAM Atomique.

The exploration program delineates success-contingent, staged exploration programs. Four stages are required to reach the point of a production decision on each of the four properties: Folkara, Faratsiho, Tranomaro, and Makay. Included within these perimeters are sites of historic uranium production by a French Agency, as well as known uranium occurrences and in some cases historic uranium resources. The exploration budget for each perimeter is upwards of $3.9 million (USD) over the course of the four stages, which culminate in the completion of a preliminary feasibility study. Work in the various perimeters has commenced on a staggered basis as a result of final Malagasy State ratification and approval of the joint venture and granting of requisite environmental permits.

To date, eight permits covering 377 carrés or an aggregate of 2356.25 square kilometres were issued to PAM Atomique to cover the Folakara zone. Located in the province of Mahajanga and Toliara, the Folokara zone is generally situated in the uraniferous sedimentary Karoo belt in Western Madagascar. It includes the zone of Morafenobe at the west-northwest and the zone of Folakara at the west.

In addition, two permits covering 27 carrés, or an aggregate of 168.75 square kilometres were issued to PAM Atomique for the Antsirabe zone. Located in the province of Antananarivo in Central Madagascar, one permit is situated in the lacustrine basin of Antsirabe approximately 180 kilometres south of Antananarivo. The other permit is in relative close proximity but is within the pegmatite fields of Vavavato.

The French Commissariat a l'Energie Atomique (CEA) exploited uranium from the regions of Folakara and Tranomaro during the period from the late 1930s through the 1950s, with systematic exploration of Madagascar for radioactive minerals being undertaken to 1966 in detailed regional studies that outlined several uraniferous areas. In 1976, the newly created OMNIS resumed exploration of the more promising CEA discoveries with technical advice and funding from the International Atomic Energy Agency and associated agencies. PAM Atomique's land position in Folakara and Antsirabe covers many known uranium occurrences and areas examined by CEA and OMNIS.

A particular area of priority for uranium exploration identified by Pan African is the Malagasy Karoo belt, which occupies a rift basin on the western side of the island and spans more than 1,400 kilometres. These Carboniferous to Jurassic sedimentary rocks were formed during the earliest stages of tectonism, that later resulted in the separation of Madagascar from Gondwana, the southern supercontinent formed of most of the southern hemisphere's land masses. The Malagasy Karoo bears similarities to the better known and mineralized Karoo basins in mainland Africa. Continental sandstones and interbedded shales predominate, and features such as fluvial beds, abundant petrified wood, and vanadium enrichments compare with features of the Colorado Plateau uranium belt. Numerous radioactive occurrences were detected by airborne and ground surveys by the CEA, which led to the discovery of showings of carnotite and other ultraviolet minerals. Madagascar's Karoo is a promising environment for sandstone-type deposits and Pan African intends to apply state-of-the-art technology and exploration techniques to define source regions, basin flow, and depositional traps with a view toward generating bona fide drill targets.

A pair of young Malagasy engineers Andrianirina Fanomezana (Fafah) Ramboasalama and Olivier Rakotomalala are among 40 local Malagasy geologists, management and technical personnel that Pan African employs to guide its operations in Madagascar.

Ramboasalama, Resident Manager for PAM Madagascar SARL and the company's Regional Manager for Southern Africa is an environmental engineer with over eight years' of experience in rural agricultural development and environmental management throughout Madagascar. He worked for three years with the Financial Company for Industry and Trade (COFIC) and was responsible for developing agri-business investment projects. In 1998 he joined CARE International in Madagascar where he was the national Program manager for their Integrated Community Development (IDP) project. Here, he helped guide CARE's work in partnership with the Wildlife Conservation Society in Masoala National Park, Madagascar's largest protected national park. He was responsible for designing and managing environmental and natural resources and community development activities. He joined Pan African shortly after its formation in 2003.

Rakotomalala, Associate Resident Manager in Madagascar and Pan African's Technical Director, is a civil engineer who ranks among Madagascar's leading GIS experts. He boasts over ten years' experience in design, implementation, and exploitation of geographic databases. These skills are invaluable in relation to Pan African's mining exploration activities. His prior experience includes a position with the National Geographic Institute of Madagascar, where he was responsible for the acquisition and management of data including surveying, aerial photography, and remote studies. He also worked for several years with CARE International, where he was responsible for designing, monitoring, and evaluating social and economic development projects. He joined Pan African in 2003 as its Technical Director and has overseen its mining claims acquisition and the management of its properties in Madagascar since that time.

The expertise and competence that Ramboasalama and Rakotomalala bring to their positions has been a huge help to Olian, who adds that their positive relations with members of government have given Pan African a leg up in the country. The company has also won kudos for its appointment of local Malagasy to such high positions of responsibility. "They run the day to day affairs of the business and as a result our company has been singled out and given accolades for its policies of true partnership with the people of Madagascar," Olian explains. "It's not just lip service," he adds. "All of our Malagasy staff, employees and consultants that have been there for over a year get stock options, they participate in the equity of the company. They are well paid and they're given a lot of responsibility. We always hire Malagasy people wherever we can."

In addition, Pan African posts a summary of its sustainability policy on its Web site, a sign that it's willing to be held accountable if it doesn't live up to the goals it sets for itself. "Maximizing shareholder value while maintaining social and environmentally responsible business practices are the basic tenets of our corporate philosophy," the site states. "We're focused on creating and sustaining a profitable business entity. We'll always do this while seeking to make life and living conditions better than when we found them. Pan African Mining Corp. is supportive of the cultures and traditions of the countries in which we chose to do business. We are socially and environmentally sensitive to the needs of the people and the environment."

While sustainable practices and mining may not always mesh with some people's views of the industry, Olian expresses a sincere concern about the conditions he's seen in Madagascar. Again, its one of the reasons he's made a commitment to hiring Malagasy staff whenever possible. The result is a smooth-running company that has good relations with the government and good prospects on the ground—and locals who know how to take care of business on local turf. "Our staff is so talented they've taken a lot of the burden off of my shoulders, so I can now be more of an administrator, which is what I was intended to be," Olian says.

What Olian has fondly called the "next great big adventure" of his life wasn't his intended destination when he graduated from university. Studies at Harvard and Princeton were preparing him for a career as a lawyer, but he soon became attracted to the business world and the intricacies of financing. Stints with Bear Stearns & Co. Inc. (he eventually became a vice-president) and Sutro & Co. Inc. in California grounded him in the world of business and he was hooked. "My entrepreneurial instincts were too strong," he says, going on to outline the string of businesses he has guided to success.

North American Scientific Inc., which he established in 1990 and which still exists, manufactures radio isotope products for environmental control and nuclear medicine. A key line of products are brachytherapy sources, or "seeds," for the treatment of prostate cancer. Based in California, North American's IPO was coordinated through Vancouver, and gave Olian connections he was able to build on for later ventures. He says it was an especially good experience from a management point of view, as it taught him lessons about building a business, raising capital, and managing people and assets. BioSource International Inc., another California company, had a similar origin. It went public with the help of a team in Vancouver and eventually rose to $36 on NASDAQ, Olian jokes "I became spoiled by the success," keenly aware that exploration doesn't always guarantee discovery in the mining business.

But, describing himself as an "opportunistic entrepreneur" and throwing out references to his work in the entertainment industry (he serves as CEO of NeoClassics Films Ltd., a privately-held, independent motion picture company and has been a senior attorney for Warner Bros. Inc. in Burbank, California and Business Affairs Exec for leading international talent agency, International Creative Management), he touts his lifelong interest in financing and company development as the rationale for his shift into mining. Though he had some investing experience with a number of junior mining companies prior to taking the helm of Pan African, the opportunity to head his own mining company came at an ideal time.

"Mining for me was just another exercise in developing a young company in a new industry," he explains. "If you can successfully launch a biotech company, a lot of the same factors are applicable." Olian's connections in Vancouver were also handy in launching Pan African, which began trading on the TSX-Venture exchange in 2004. While the financing contacts were developed during his years in biotech, Canada was the place he knew he had to come to if he wanted to launch a successful mining company.

"The Toronto Stock Exchange and the TSX-Venture exchange, in my opinion, are the best mining markets in the world, the best sources of mining finance," he says. "I felt that it would be the most appropriate to get the capital here in Canada, and I had a lot of connections in Vancouver." Moreover, the high degree of regulation governing the TSX exchanges promised credibility, something one can't have too much of in a heady market where investors want to see a quick return from hot commodities.

The change of pace from biotech to mining has been an especially important shift for Olian, who enjoys the opportunity to learn. "It's a challenge for me, which I really enjoy; I like the stimulation of learning about new industries. I'm very excited about the mining business. But I think it's clear that in any successful business, there are certain fundamental parameters that have to be present. Certain standard rules of operation are required to make any business a success—integrity, economy, hiring really top quality people, having a very good business model, and constant vigilance."

Working in the developing economy of Madagascar is also a challenge. Olian has found the assistance and support of Chris Dunston, and his brother Lee, invaluable in helping Pan African develop relationships with the Malagasy government as well as local workers, and in ensuring the smooth establishment of the company in Madagascar for the benefit of stakeholders all round. "Chris helped introduce us to a lot of the government people and laid the groundwork for us from an infrastructure and administrative standpoint," Olian says.

The impressive background Olian brings to his position is complemented by a team of executives and corporate directors with their own unique skills and expertise. CFO Limor Rubin, for example, brings eleven years' experience in public accounting practice to the company. She served as accounting manager for KPMG and as a senior accountant with both KPMG and Ernst & Young. Rubin is a member of the Institute of Chartered Accountants of British Columbia. She is also a qualified Chartered Accountant (Canada), a Certified Public Accountant (Israel), and holds a Bachelor in Accounting and Economics from Tel-Aviv University. In addition to her experience in the mining industry, she has experience with the financial services industry, mutual funds, insurance, and real estate as well as in the public sector.

Gregory Sparks, Director, is a registered professional engineer with over 25 years mining experience. Sparks serves fulltime as senior Vice President of Pan African's Latin American sister company Sacre-Coeur Minerals Ltd., overseeing exploration and mining activities. Sparks was formerly Vice-President, Development for Echo Bay Mines Ltd. for eight years in the 1990's.

Ardito Martohardjono is also Vice-President of Laboratory services and has 15 years of progressive experience in the mineral exploration industry. His experience has focused on sample preparation prior to chemical analysis, and assaying of geological samples. He has served as a lab technician and supervisor with the two leading geochemical laboratories in Canada and the United States. During the past five years he has served as sample preparation supervisor for Bondar Clegg Canada Ltd. in Vancouver and as branch manager of Bondar Clegg's laboratory in Fairbanks, Alaska, as well as coordinator of the prep client services office and sample preparation supervisor for Chemex Labs Ltd. (now ALS Chemex). Among his many responsibilities was ensuring compliance with ISO quality control standards.

Director Dr. Edward Schiller serves as senior geological consultant for Pan African's diamond and gemstone programs. Dr. Schiller brings over 30 years experience in mineral exploration, project management, acquisitions, financing, joint venture negotiations and corporate governance to the Company. Dr. Schiller is a former director of Dia Met Minerals Ltd., and is best known for supervising the drilling which led to discovery of the first diamond-bearing kimberlite at Pointe Lake in 1991 (now part of the Ekati Mine production). He has lived and worked in Canada, the United States, England, Australia, Brazil, Columbia, and has conducted mineral exploration projects in several South and Central American, African and South East Asian countries, including Madagascar. A veteran geologist, he writes for several Canadian and international magazines on mining and mineral exploration and maintains his own consulting practice.

In addition to directors such as Schiller and Martohardjono, Pan African enjoys the expertise of an extensive geological team of nine people that provides valuable advice during the selection of new exploration sites. The uranium program is headed by Dr. Reinhard Ramdohr, a German geologist with over 30 years of international experience covering uranium, gold, base metals, and gemstones. He has overseen grassroots mineral exploration programs in diverse settings on nearly all continents. His African experience is extensive, covering projects in Ethiopia, Eritrea, Ghana, Tanzania, Senegal, Central African Republic, and Madagascar. In Madagascar, where he has worked on several occasions, he discovered a significant sapphire deposit.

The diverse expertise is appropriate to a company with as many diverse mineral interests as Pan African has. Olian views his role chiefly as the coordinator of the company, making sure it stays on track as it strives to find deposits that will deliver an economic pay-off. "I'm smart enough to recognize that I need to be an administrator and flag-waver, and I try to attract the best talent I can to be the on-the-ground professionals," he says. "I like to let our geologists do their jobs with as little interference as possible," he adds. "I do not micromanage them."

Nevertheless, Olian keeps close tabs on expenses and exploration programs and makes a point of suggesting changes in strategy from time to time. After all, Ramdohr and the others delving into Madagascar's uranium fields may be working in a rich geological environment, but hardly one that's a picnic.

The area encompassed by Pan Africa's permits feature a wide variety of topographical and geological diversity, from mountainous regions believed to bear significant hard-rock deposits to areas of low relief featuring numerous river systems and alluvial deposits. Some areas are arid, semi-desert environments that are nearly treeless, while others are rich, tropical rainforest. The diversity of terrain is mirrored in the diversity of access methods. Some areas are easily accessible by road, river or airplane, while other areas are remote with very difficult access, even by 4-wheel drive, helicopter, or on foot.

But, as Olian emphasizes, the prospect of a pay-off from Pan African's uranium exploration is spurring an ambitious drilling program. Olian's ultimate goal is to establish a series of regional clusters that will justify a feasibility study for establishing a mine. That's two or three years away yet, Olian estimates, which means a busy period lies ahead for the company. While nothing found in Madagascar yet rivals the richest, and much of which is currently in production around the world, Olian said Pan African's properties on the island offer many deposits that are richer than most of what is currently in production. He also notes people mine gold for far less cash per ton than what uranium currently fetches.

In the case of the Tranomaro uranium property, results from Pan African's initial drill holes suggested a deposit with high grade uranium in some cases approaching 10 lbs/tonne of U308, or about 0.4%. Typical grades encountered are 1.5 - 7 lbs/tonne U308. Olian considers the early results extremely encouraging, noting that many mines are currently extracting ore grading at less than 0.1 per cent. "We think this certainly has the appearance of being economic, subject to our delineating a big enough mass of material," he says. Olian was enthusiastic enough about the initial results, announced in June 2007, that Pan African commenced a formal diamond drilling program the following month.

A heavy Atlas-Copco CS-14 diamond drill rig was brought on site and, by the beginning of September, nine holes had been completed to depths of between 70 and 130 metres at the old French Mine Number 37 in the Tranomaro area. A series of core samples were sent to Vancouver for assay, and results are expected over the course of the fall.

Further drilling is planned in the Tranomaro area and Pan African is investigating more uranothorianite prospects north of Tranomaro at old French Mines Numbers 52 and 52N, where shallow scout core drills are developing deep drill targets. Mines 53, 54, and 55 are also under investigation. Pan African has also identified targets at old Mines 49/50 and 26.

Ironically, the discovery of a significant uranium deposit may be overshadowed by another noteworthy result of Pan African's exploration efforts on Madagascar. During the course of his work with Pan African, Olian has become increasingly interested in the prospect of extracting thorium from the ore bodies Pan African is pursuing. The idea is guided by the end purpose of the exploration—the discovery of uranium that will provide an alternative fuel source for power plants.

"I share the belief of a large portion of the environmentally sensitive community right now that uranium really is a new green source of energy that has proven itself to be safe and reliable when treated properly and with modern technology. I think uranium, as a potential fuel source of the future, holds out great hope when properly managed and controlled" he states.

But, as Olian explains, thorium has good potential to become another alternative fuel source. India and Norway are already considering using thorium as a cheaper, less dangerous nuclear fuel than uranium. While equally powerful, thorium doesn't produce the runaway chain reaction which has the potential of leading to a meltdown in a uranium-fuelled reactor. Moreover, thorium reactors produce a fraction of the waste created by uranium reactors. And, perhaps best of all in a security-conscious world, thorium can't be enriched to provide weapons-grade radioactive material. However, it does yield just as much power as uranium when processed correctly.

This is what prompted Statkraft, Norway's state-owned energy company to pursue plans for a thorium-fuelled nuclear reactor. It's Norway's plans that encourage Olian.

Thorium has typically been removed from ore and disposed of, which is exactly what would normally happen at any mines Pan African's exploration work leads to. But there is now a simple, effective chemical process that can extract thorium from ore and render it useable for nuclear fuel. Rather than disposal being an added cost that increases the price of uranium extraction, recovering thorium could pay for itself. Indeed, Olian believes the grades in the ore on Madagascar are so rich that there could be a significant reward to be had. Minerals, such as thorianite contain significant amounts of thorium.

Indeed, the world's largest thorianite crystal was found on Madagascar—a six-centimetre crystal weighing nearly five pounds.

In short, extracting thorium from the uranothorianite ore would be economically feasible if a market develops. "Pan African may be sitting on the free world's largest deposit of thorium," Olian says, his enthusiasm shining through. "The grades are so rich that the added cost factor would appear to be easily covered by the high-grade uranium that we're finding." Pan African won't act on Olian's vision until there is a market for thorium but in the meantime, Olian expects marketing of the metal to increase as a means of building awareness.

The irony, of course, is that if thorium is accepted as a cleaner, safer—better—nuclear fuel than uranium, Pan African could be well-positioned to meet the new demand. "It ultimately may be the bigger market for us," Olian says. "That would be our trump card, if thorium became the next new alternative energy source."

In the meantime, Pan African intends to grow along with its various projects. Regardless of how they progress, Olian feels Pan African has the expertise and the connections to make a go of any deposit its geological teams discover. "We're not wedded to the idea that we have to be the producer," Olian says. "But we want to develop the in-house capability."

In addition to Olian's own extensive experience and connections in the world of corporate finance, its geological teams have connections to the major gold and diamond players. The ambition and in-house capabilities are one reason Canada's eResearch investment analysis firm went bullish on Pan African in July 2005. Noting that the company was one of the first to enter Madagascar in the wake of mining reforms, analyst Barbara Thomae noted the diversity of its holdings, pointing especially to the recent joint-venture arrangement with the Madagascar government that allowed its uranium exploration activities to commence.

"We believe that Pan African Mining's value will be unleashed as it uses modern methods to gain better insight into the full potential of its holdings," she writes in her report. Thomae rated Pan African a "Speculative Buy" at the time and set a target price over the following twelve months of $1.65 a share. The stock now trades in excess of $2 a share.

"We feel we have the capabilities to put together the proper teams to bring our projects to fruition and into production," Olian says. "And that's our intent unless and until we are approached on a particular project with an offer we can't refuse."

Silver Spruce Resources Inc.

"The Big Land" is what those who are familiar with its vast reaches call it. It's Labrador to the rest of the world, tucked away to the northeast of Quebec, Canada's biggest province, its rugged face turned to the wind-whipped North Atlantic Ocean. With a territory of 269,073 square kilometres for a population of just 27,000 people, Labrador is easily one of the most sparsely populated regions in Canada.

Dreams and schemes for projects designed to tap the wealth of natural resources have abounded through the years, as hydro projects and mining ventures of all manner have proven only too well.

Indeed, Silver Spruce Resources Inc. (SSE:TSX-V) might have had a better start in the exploration business if one of those ambitious ventures hadn't ended the mineral resource boom that was just beginning in the 1990s on the backs of the September 1993 Voisey's Bay nickel discovery.

Created in April 1996 by celebrated Labrador entrepreneur Lloyd Hillier, Silver Spruce was initially known as First Labrador Acquisitions Inc. with a mission to acquire properties for metal and mineral exploration. A junior resource company based in Goose Bay, First Labrador was listed on the now-defunct Alberta Stock Exchange and had some initial success searching for zinc and other metals in Central Newfoundland on properties previously worked by Noranda Exploration Co. Ltd. and others.

But its initial intentions were largely stymied by the boondoggle associated with the claims Cartaway Resources Corp. made for its holdings in the Voisey's Bay area 350 kilometres north of Goose Bay.

A newcomer to the mining industry, Cartaway announced in June 1995 that it had acquired several claims in the vicinity of Voisey's Bay, where Inco is currently proceeding with development of a major nickel mine. The following spring, Cartaway issued a handful of press releases indicating that its initial test results in the area had been encouraging, with good showings of copper and nickel. The only trouble was, the results weren't quite so promising, and fallout was even less so for exploration companies working around Voisey's Bay.

The chill this threw on exploration companies working in the Voisey Bay region was immediate, and First Labrador was caught in the ubiquitous freeze.

"The juniors from that point on couldn't raise any money. So First Labrador, which had acquired some ground up there really couldn't do much with it at that point," recalls Peter Dimmell, Chair of the Newfoundland and Labrador Chamber of Mineral Resources and now vice-president, exploration, for Silver Spruce.

The downturn did nothing to diminish Hillier's enthusiasm for mining prospects in his home province, however. Hillier knew he would be back for another run.

Try as he might to locate suitable opportunities, the luck just wasn't there for a decade. No matter the ground First Labrador acquired, the prospects Hillier and his technical teams scouted or the research brought to bear on the regions, nothing bore fruit.

"They really didn't have a lot of luck getting much going," Dimmell says. He recalls with a laugh the enthusiasm with which Hillier bought a diamond drill for a property he was prospecting on the Burin Peninsula of Newfoundland. He had great hopes – but the drill yielded nothing.

While tech companies set new records for valuations and IPOs for knowledge-based companies blossomed, resource-based companies and junior mining ventures languished in the shadows. Many looked at other industries for ways to remake themselves, eager to get on with business in a world that had gone upside-down.

"You really couldn't do a lot. It was the time of the dot-com boom. You couldn't raise any money, you couldn't do anything," Dimmell said. "[Lloyd and his team] looked at a bunch of different alternatives for the company. They looked at waste to energy stuff, they looked at a couple of other different projects, but nothing ever seemed to be something you could take and make something of."

Uranium, a heavy white metal then trading somewhere below US$10 a pound in the early 2000s, was the last thing anyone was thinking of.

While the Central Mineral Belt in Labrador was heavily prospected by British Newfoundland Exploration Ltd. in the 1950s through the 1970s, it had come to hold more interest for its characteristic copper deposits than its more elusive uranium. When copper prices fell in the 1970s, British Newfoundland's interest tailed off and it let its claims lapse in 1983. It wasn't of much interest to anyone until uranium prices started making significant gains about three years ago.

One of the first companies in was St. John's, Newfoundland-based Altius Minerals Corp., which was staking claims in the area as early as 2001, and subsequently put the Cental Mineral belt on the map through the formation of the TSX-Venture-listed company Aurora Energy Resources Inc. Altius had an idea that the area around Michelin Lake and other parts of the Central Mineral Belt bore a geological resemblance to the Olympic Dam mineralization in Australia, which lays claim to being the largest uranium mine in the world.

Drilling activity by Aurora led to estimates of the uranium deposits in the Michelin Lake area of the Central Mineral Belt rising from a paltry 12 million pounds to in excess of 100 million pounds. Aurora is now pursuing a feasibility study regarding the extraction of the uranium, which stands to net Altius a solid royalty should extraction proceed.

Crosshair Exploration and Mining Corp. followed, optioning the Moran Lake deposit. Then Santoy Resources Ltd. stepped in and the Labrador uranium rush began gathering momentum. The appreciation in uranium prices added fuel to the movement, and suddenly uranium became a priority for companies wishing to be active in Labrador.

The rush was on.

By this time, Silver Spruce Resources had been adopted as the new name of First Labrador Resources. The company was still focused on exploration, but it rechristened itself after one of Labrador's most distinctive species of trees in October 2004 with high hopes for a new era.

The growing interest in uranium wasn't lost to Hillier, who had been scouting properties in the Central Mineral Belt with well-known Newfoundland prospector Alex Turpin since 2005. The attention of Hillier and Turpin was focused on the Seal Lake area of central Labrador just west of the Central Mineral Belt. They were looking for copper, not uranium, but the growing excitement was hard to ignore.

"All this activity is going on just east of the Seal Lake area where we had all these claims staked for copper," Dimmell says. "Lloyd sees all this, sees the things going on, and says there's no reason why we can't be involved in this, too."

Silver Spruce began staking ground in the Central Mineral Belt in January 2006, adjacent to where there were good showings from the other companies. It eventually picked up all the major showings in the area and most of the ground, filling in what it didn't already have. In January 2006, the company announced that it had acquired 4,963 claims across approximately 1,240 square kilometres. Silver Spruce then optioned to Universal Uranium Ltd. of Vancouver, the right to earn a majority interest in these properties. Under the terms of the agreement, Universal could earn a 60 percent interest by spending $2 million on exploration over a three-year period with a minimum of $500,000 the first year and $750,000 each of the following two years. Silver Spruce was to be the operator during the earn-in period.

The following September, Silver Spruce announced that it had acquired an additional 600 claims tied on to the north and west of the original Central Mineral Belt block. These new claims were subject to the Silver Spruce/Universal Uranium joint-venture agreement. The new properties are proximal to the Michelin, Moran Lake and other uranium showings being explored by Aurora Energy, Crosshair Exploration and Mining and Mega Uranium Ltd.

Silver Spruce now claims the second-largest land position in the Central Mineral Belt, and today holds 15,000 claims across 32,000 square kilometres of prime central Labrador mining country. In addition, Silver Spruce has rights to four other properties where it is prospecting for uranium.

The Double Mer property in eastern Labrador, approximately 110 kilometres east of Happy Valley-Goose Bay, consists of 758 claims on 190 square kilometres in a single contiguous block. The claims were acquired in an arm's-length deal with Alex Turpin and have the potential to yield strong uranium deposits from lake sediment anomalies located by the Newfoundland and Labrador government.

The Mount Benedict property consists of 1,048 claims across 262 square kilometres in five separate blocks. Located near Makkovick approximately 180 kilometres northeast of Happy Valley-Goose Bay, the property more than doubled in size in August 2006 when Silver Spruce acquired more than half of these claims, subject to an option agreement with an independent geological consultant. These properties cover uranium in lake sediment anomalies located by the Newfoundland and Labrador government.

The wholly owned Straits project consists of 800 claims across 200 square kilometres in one contiguous block located in the area of Barge Bay-Henley Harbour, approximately 300 kilometres southeast of Happy Valley-Goose Bay. Like the other deposits, the claims cover uranium in lake sediment anomalies. These anomalies were located by the Geological Survey of Canada.

Beyond Labrador, Silver Spruce signed a letter of intent in April 2007 with Quebec-based Azimut Exploration Inc., a TSX-Venture listed company, regarding the Hudson Bay uranium property in northern Quebec. The letter gives Silver Spruce rights to acquire a 50 percent interest in the project from Azimut over a five-year period together with an additional 15 percent interest upon delivery of a bankable feasibility study.

The Hudson Bay property, located near Hudson Bay within 40 kilometres of Umiujaq, consists of three blocks totalling 537 claims with a surface area of 253 square kilometres. This includes eight recently staked claims for which confirmation is pending from the Ministry of Natural Resources and Wildlife of Quebec. Like the Labrador properties, the Hudson Bay property features traces of uranium in lake sediment anomalies. In its favour, uranium values of up to 1.31% have been reported in rock samples in an area where exploration is now prescribed. At Hudson Bay, the target type is a large intrusion-related uranium deposit amenable to open pit mining.

In addition to its uranium properties, Silver Spruce holds interest in two gold properties with one located on the Burin Peninsula of Newfoundland and the other in Mexico's Chihuahua State. These properties will likely be spun off into a separate company in due course.

Hillier wasted no time in reorienting Silver Spruce and getting on with the business of uranium mining. Setting aside his goal of copper, he initiated airborne surveys of the Silver Spruce's Central Mineral Belt properties in summer 2006, and undertook radiometric studies to identify the areas with the greatest radioactivity. Working with geologist and company director Ted Urquhart, an internationally recognized professional currently based in Chile, Hillier pursued further studies to determine the source of the radiation, whether from potassium (not usually a big issue), thorium or uranium.

"These guys went in and they spent about two weeks with a helicopter and scintillometers – these radiation detectors – they would figure out where the radiometric anomaly was and they'd fly the helicopter to that co-ordinate area," Dimmell explains. "They'd get out and walk around with the scintillometers and see if they could find out where the source was. They didn't have a lot of success in most of the areas."

That is except for one. Out of 17 targets across Silver Spruce's Central Mineral Belt properties, CMB Northwest No. 2 showed the most promise – the so-called 'Two-Time' zone.

Silver Spruce's second major foray into central Labrador wasn't without its share of curious repetitions. The name for the zone was originally chosen because Hillier returned to the property to locate the potential deposit, as the original team that went in had been about 600 feet off because the coordinates on the GPS locator system they were using were off. Not one to take no for an answer, Hillier went back himself and found the radioactive outcrop that the air surveys pointed to. There were good values in the outcrop and in rocks scattered about the area, and the area also held what seemed to be the source of the CMB No. 1 deposit, which Crosshair is keen on realizing.

On a later occasion, Hillier had to go back two kilometres and fetch samples taken from the Two-Time deposit, which everyone thought someone else was carrying with them.

Then the hard-won samples were lost when they were sent to the lab in Ontario via Priority Post.

Hillier and Dimmell went back and got more samples, shipping them off to the lab again.

It was the third time something had to be done twice, and the name for the Two-Time zone stuck.

The backtracking in the Two Time zone was uncharacteristic of Hillier, a seasoned entrepreneur more used to acting with intent than having to repeat himself. Still, he has proven his mettle on more than one occasion, and in many more trying circumstances. The decade of disappointments that preceded Silver Spruce's prospects in the Central Mineral Belt were dues paid out of the reserve of fortitude that helped Hillier build a significant business empire in Labrador.

Named Newfoundland & Labrador's Entrepreneur of the Year in 2001, Hillier got his start in business in the retail business, which remains a mainstay of his holdings. He was also an electrician at CFB Goose Bay for many years, and remains active in the construction business through Hillier's Trades Ltd., which provides hardware and supplies to communities in Labrador. It is so successful that it once received a buy-out offer from a national hardware chain. Hillier Trades also owns and operates tractor trailers, a construction division, and apartments in Goose Bay. Hillier also operates Hotel North, considered by many the best hotel in the area. Of note, Hillier's wife, Judy, manages this popular hotel and its restaurant.

Well known in Labrador, Hillier enjoys a larger-than-life persona among his business associates and even casual acquaintances, who describe him variously as a "big character" who is "very much his own man."

Dimmell points to the fact that Hillier was the one who pushed for drilling on Silver Spruce's Two-Time property in December, a month when most other companies have pulled out on account of the weather.

"Typical entrepreneur!" Dimmell says.

On another occasion, Hillier responded to rising freight rates on supplies being brought in for his hardware stores by buying his own vessel for transporting them. He picked up the boat in Halifax and brought it back to Labrador himself, though 'master mariner' isn't among his qualifications.

He's now applying that same grit and determination to his uranium venture. And, he expects the executives who join him to have the same aggressive attitude to propel the company forward. Not surprisingly, he hasn't settled for second-class.

Dimmell's own background spans four decades in the mining sector, both in North America and overseas. A geologist and a prospector, Dimmell is the immediate past president of the Prospectors and Developers Association of Canada (PDAC), a director and the chairman of the Newfoundland and Labrador Chamber of Mineral Resources, and a councillor and a member of the Geological Association of Canada, a member of the Canadian Institute of Mining, Metallurgy and Petroleum. He is also an associate member of the Association of Applied Geochemists. He currently serves as a director of four other public companies: Dragon Capital Corp., Linear Gold Corp., Pele Mountain Resources Inc., and VVC Exploration Corp. Prior to joining Silver Spruce in 2005, he was with Silver Spruce's rival in the Labrador uranium field, Crosshair Exploration.

Silver Spruce chief financial officer Gordon Barnhill, who also serves as vice-president, corporate affairs joined Silver Spruce following several years heading a company providing a variety of consulting services including management consulting, capital research, business valuations, deal structuring and investment strategies. From 1973 to 1997, Barnhill was a senior commercial lending officer with Canada's largest banking institution.

In addition, Silver Spruce's directors include Dr. George Findlay, a graduate of prestigious Rothesay Collegiate School in Rothesay, New Brunswick, which has been attended by other luminaries of Atlantic Canada's business world such as the children of business magnate K.C.

Irving. Findlay studied science and dentistry at Dalhousie University in Halifax, and in 1977 established his own dentistry practice in Bath, New Brunswick. Following his retirement in 2006, Findlay studied towards an MBA at England's University of Liverpool, graduating in 2007. He subsequently completed the directors' course at Simon Fraser University in British Columbia and obtained his prospector's license by completing a course in prospecting offered through the New Brunswick Department of Natural Resources. A long-time investor in the resource sector (his first investment was in Brunswick Mining and Smelting Corp. in 1964), Dr. Findlay has been an ardent student of public market activity with a particular focus on mining issues.

The breadth of experience overseeing Silver Spruce's business activities is complemented not only by Findlay's recent licensing as a prospector, but by a technical team that includes Guy MacGillivray, Silver Spruce's senior geologist. MacGillivray brings extensive experience in the exploration and mining industry to Silver Spruce. His resume includes work on uranium projects with Eldorado Nuclear Ltd. and Shell Canada, as well as two years working with Scorpio Mining Corp. on the Nuestra Senora Project in the Sinaloa, Mexico. At Sinaloa, he was responsible for the construction and maintenance of the drill hole and underground sampling database, deposit modeling, and internal resource estimates. This work was instrumental in helping to advance the understanding of the ore controls and contributed to the discovery of a number of new ore bodies, including the Hoag and September 9th zones, which helped advance the project to a production decision.

A graduate of St. Francis Xavier University in Antigonish, Nova Scotia and a member of the Association of Professional Engineers and Geoscientists of Newfoundland since 1990, MacGillivray has spent 29 years working in the exploration and mining industry throughout North America. He worked for over 25 years as an exploration geologist for companies such as Rio Algom Ltd., B.P. Selco and Teck Ltd. and was involved with exploration and development work at the East Kemptville tin deposit, the Hope Brook gold deposit and the Voisey's Bay copper-nickel deposits. MacGillivray is also credited with the discovery of the White Rock silica kaolin deposit in southwestern Nova Scotia.

An active participant in the expedition that led to the identification of Silver Spruce's Two-Time zone in the Central Mineral Belt, director Ted Urquhart brings an international clout to the company. Though he maintains a residence in Bedford, Nova Scotia, his business base is in Santiago, Chile, where he operates a geophysical consultancy. Backed by more than 25 years of international experience in geophysical surveying, he has overseen radiometric, magnetic and electromagnetic surveys, both with the Geological Survey of Canada (GSC) and with several private companies. Most recently, he ran his own survey company, High-Sense Geophysics, prior to its merger with several other companies in 2000 to form the international consulting firm Fugro Airborne Surveys Corp.

Additionally, Urquhart is no stranger to uranium. In the late 1970s, he worked on a number of uranium exploration projects including Gulf International Minerals Ltd.'s work in the Great Bear Lake area of the Northwest Territories. He has also worked with the Comisión Nacional de Energía Atómica of Argentina, France's Cogema (now Areva NC) in Mali, and with the Geological Survey of Canada on a five-year airborne survey that developed new processing techniques for radiometric surveying. From 1988 to 1993, he was a consultant to the International Atomic Energy Agency (IAEA), specializing in uranium surveying, interpretation and follow-up.

Most recently, Urquhart was an integral part of the exploration team that completed 18,600 line kilometres of airborne, helicopter-supported radiometric and magnetic surveys for uranium over the company's 8,600 claims on the wholly owned and joint ventured projects in Labrador in 2006. During this time he served as the geophysical consultant who supervised and evaluated the data produced.

While Silver Spruce has attracted a fair amount of business and technical expertise, it has also paid attention to the communities in which it operates. A sensible business operator, Hillier is aware of the importance of building relationships with the communities in which he operates. Approximately 70 percent of the 60 people Silver Spruce employs are Labrador residents, and half are First Nations peoples, either Innu or Inuit. Silver Spruce has appointed an aboriginal advisory committee to address the concerns of First Nations in Labrador, which some would have deemed a homeland for the Innu.

Four people, all of aboriginal descent and all residents of Labrador, serve on the board. They include Marcel Ashini, George Rich, Ernest McLean and Ed Montague. All are active in their communities and in aboriginal affairs.

Marcel Ashini is the development officer for the town of Sheshatshui, and works directly with the Chief of the Sheshatshui band. George Rich, a former candidate for the leadership of the Innu nation, lives in the mainly-Innu community of Natuashish on the coast of Labrador where he serves as a member of Natuashish's Child, Youth and Family Services committee. Happy Valley-Goose Bay resident Ernest McLean is a former Minister of Labrador and Aboriginal Affairs with the Newfoundland and Labrador provincial government and is a member of the Nunatsiavut government, a regional governing body for Labrador Inuit.

Ed Montague, a geologist, is the province's former representative of the Department of Natural Resources in Labrador West. He resides in Labrador City and serves alongside McLean as a member of the Nunatsiavut government. Equally significant for Silver Spruce, Montague has spent 25 years with mining and engineering companies such as Bechtel Corp., Placer Dome Inc. and United Keno Hill Mines Ltd., working in Canada and overseas. He is a graduate in geology from Memorial University of Newfoundland and a professional geoscientist registered with the Association of Professional Engineers and Geoscientists of Newfoundland and Labrador.

But all the management expertise and technical knowledge is pointless without the exercise of acute business acumen. Silver Spruce is approaching a point where it needs to make some hard decisions about its properties if it's going to have anything to show for its efforts.

"We're going to have to start dropping some of this ground because it's not all prospective," Dimmell acknowledges. "We're going to have to start reducing some of our ground because it's just getting very expensive to hold it. Obviously, there's no advantage in holding ground unless there's some reason to hold it."

While Silver Spruce has made steady progress over the past two years, it will have to make further progress in the year ahead. Dimmell knows what has to be done: Silver Spruce's property territories will be streamlined, some will be relinquished and major efforts will be put into identifying new showings on the ground that remains.

While the Mount Benedict, Double Mer and Straits projects have a lot of potential, the company's focus will undoubtedly come to bear most keenly on the Central Mineral Belt where two drills are planned to run through the winter of 2007-2008.

It's the kind of environment in which Hillier thrives.

While Dimmell's background as a geologist believes in pursuing a slow and steady approach on potential deposits, Hillier has the entrepreneur's attitude that requires things to be done, results to be got and success to be had. Keen on doing the drilling and seeing the results, his drive led Silver

Spruce to undertake the basic work in autumn 2006 that would establish a baseline for future exploration activities.

The focus, of course, was the Two-Time zone in the Central Mineral belt. Other companies were shutting down, including Crosshair Exploration, which had a drill available (competition doesn't mean Labrador's juniors don't help one another out). The sampling led to uranium deposits being identified at four out of five sites. Moreover, unlike the initial surface readings, the indications were stronger and better than what Silver Spruce had encountered to date.

"We got some fairly significant values out of the drilling," Dimmell says.

Though the values weren't exceptionally high, they were enough of an encouragement that work continued through the winter, resuming in late January and continuing into early February. The determination paid off.

On March 1, 2007 Silver Spruce announced discovery of two-pound rock from the drilling Hiller had so eagerly undertaken. The discovery was the highest grade Silver Spruce had yet identified – 0.11 percent – in a 30-metre mineralization located no more than 200 metres below the surface.

"We are very excited by this new discovery," Hillier stated at the time. "These results further indicate the potential presence of a large economic uranium deposit on our ground."

The discovery gave Silver Spruce's share price a boost. Its stock was soon trading at over $2 a share, and it has largely remained above $1 for the remainder of the year, compared to less than 60 cents for most of its previous history.

But the drilling continued, through spring breakup, the summer, and into autumn. Silver Spruce committed $4 million to exploration work in 2007, with plenty of ground follow-up to sites identified in 2006.

Silver Spruce has now sunk more than 20 holes, outlining a zone of upwards of 450 metres with almost 200 metres of mineralization. Though the more recent drill holes have yielded lower grades than some of the initial holes, Dimmell said it doesn't diminish the fact that there are substantial grades within the system and some grades that are even higher than expected.

Other teams are following up leads at other claims, with 2007 proving to be the busiest year Silver Spruce has yet seen. The Double Mer property is seeing follow up on airborne reconnaissance completed in summer 2007, while the results of 2006 airborne studies at the Mount Benedict and Straits projects are also being followed up.

But will it work out?

It's a big question, but one that may have a more attractive answer given that everything Silver Spruce has been finding to date has been through surface exploration and relatively shallow drill holes.

"Everything we have found to date – we would probably have 50, 60, 70 showings – every one of those have been found by prospecting on the surface. And that's the difference," Dimmell said. "What we're doing is hiring guys with a two-week prospecting course, giving them a scintillometer and they're going out and finding us new showings." It means a cheaper exploration cost than, say, what takes place in the Athabasca Basin of northewestern Saskatchewan.

"They're getting very high grades in the Athabasca Basin but you can't go underground in those mines. It's all robotic mining," Dimmell adds. "They've got stuff that's probably worth $36,000 a tonne, but the cost of mining is probably $10,000 a tonne."

Cheaper exploration and extraction costs don't necessarily mean a more economical deposit, however. That depends on the market price for the metal, and the size of the deposit.

Its initial results in the Two-Time zone have prompted Silver Spruce to pursue a resource estimate on its findings.

"When you do a resource estimate, it does mean that it's economic," Dimmell explains. "It means that it's a resource, and it may or may not be economic depending on the conditions, which is the price, the infrastructure costs, the type of mining – that type of thing."

The good news from Dimmell's point of view is that uranium prices still seem to be on the upswing, and even if prices drop below US$100 a pound, a sizeable deposit of the kind announced in the Two-Time zone in March 2007 could be considered economic.

"I think you've got a very good chance of making a mine if it's bulk mineable or open-pittable," Dimmell says, adding that he doesn't believe uranium has to go to $150 a pound in order to be profitable.

But he emphasizes that Silver Spruce isn't anywhere close to becoming a uranium producer in the near future, even if it had plans to follow this route on making a discovery. "We are still an exploration stage company," he says. "We are not even into the advanced exploration stage at the moment."

But the signs for a discovery have been positive to date, and that buoys Dimmell's hopes.

Besides, if someone else beats Silver Spruce to the punch and announces a major discovery and plans for a producing mine, the appeal of any deposit Silver Spruce announces will be that much more attractive. Any company's deposit, he thinks would bring the majors calling.

The nearby Michelin deposit, for example, could be in production by 2011, which would be a huge benefit for the economics of any discovery on Silver Spruce's claims. The infrastructure would be put in place – whether road, port or airport, and a mill would be constructed.

And many think that scenario is closer to fact than fiction on the exploration being done.

"Between ourselves and the other companies working up there, I think there's going to be substantial resources of uranium defined, some of which will be economic under present conditions," Dimmell said. "If we can come up with 10 or 20 million pounds of uranium as a resource we're on the road to being a producer."

For now, however, Silver Spruce will continue with its exploration, confident that glory awaits. Third-phase drilling has found further uranium mineralization in the 0.1 per cent range, boosting the anticipation Hillier and his team have felt.

"We started off as a little exploration company with, really, just proximity ground, and now we're developing into a company that will hopefully have the uranium resource pretty soon," Dimmell says. "While it doesn't indicate economic viability, what it indicates is at least you've got something better than just an intersection and a drill hole. And of course, that's what people want to see."

Strathmore Minerals Corp.

When Devinder Randhawa thinks back on the history of Strathmore Minerals Corp. (TSX-STM.V) and the uranium asset acquisitions his company has made, he notes that "a little contrarian thinking can go a long way." Thanks to that, a solid business strategy and an experienced management team, Strathmore has become a major player in the North American uranium industry. Within the next 3 years it will also likely become one of the newest uranium suppliers in the United States when the Gas Hills project the company has been developing in Wyoming reaches production.

Dev's decade-long progress to that achievement was neither simple nor easy. He says it has involved the support and advice of many people and the dedication of a particular management partner who has invaluable uranium industry experience. "You can have an idea but it's really important to make sure you listen to smarter people. I was advised by some industry legends and if those guys didn't say they were onside, I'm not sure I'd be where I am."

Strathmore's evolution began in 1996. At a Las Vegas gold show, Rick Rule, the intellectual force behind Global Resource Investments, suggested to Dev that he enter the North American uranium space.

Before the investor event, another broker named Richard Newberry told Dev he'd had an interesting conversation with a uranium industry old timer. The prospector was looking for an investor and Newberry wondered if Dev was interested. The unknowns of that risk left Dev uncertain, so he hesitated. His uranium industry experience at that time was negligible and the common industry opinion was to avoid uranium investments. Major mining companies that had maintained huge uranium exploration budgets were trying to strip the resource from their asset portfolio.

"But, in my view, Rick is one of the best brokers in the resource sector in North America and especially in a bear market. He's always ahead of the curve. Rick thinks outside of the box and he's a very intelligent guy," says Dev. "He said look at uranium and I'll support you if you do. I listened to him and decided to get more aggressive."

Dev created Strathmore and soon got an option to buy some pounds in the ground in New Mexico. At that time Strathmore's stock was floating in the 30 to 50 cents range. With a financing organized by Steve Khan (now Strathmore's executive vice-president) the stock eventually peaked at over $2 a share. "I was on a steep learning curve as far as uranium was concerned though," says Dev. "The deal I was looking for at the time was optioning land for a dollar a pound, and towards 1998 I met Dave Miller. He asked me why I was spending that much when he could go and get it for pennies a pound. Naturally I listened to him."

David Miller had entered the uranium industry in 1976 as a 23 year-old geology grad when he went to work for Pathfinder Mines Corporation, a subsidiary of General Electric. He worked shoulder-to-shoulder with industry veterans.

"These were the people that founded the uranium industry back in the 1950s and 1960s. That's where I got my roots. It was from the people who invented the industry," Miller says.

In 1982 Cogema purchased Pathfinder and Miller worked for the international uranium giant through two decades. "I've been through the uranium cycle. I entered it in the last boom and rode it through the trough. I was one of the survivors," he says, and he'd become a minerals industry expert in exploration, acquisition and operations. His primary focus had been on uranium, coal bed methane and gold, and he'd served as Cogema's chief geologist for in-situ operations in the U.S.

Miller joined Strathmore to help with new acquisitions but over the next two years the spot price of uranium tumbled from the $12 range to its spot price low of $7.23 per pound as of January 2001. Investor confidence disappeared. "We ran out of money," Dev recalls, "so Dave went back to consulting work and I did some other things." Those 'other things' turned out to be profitable endeavors and that profitability was critical to Strathmore's future survival.

Dev may have been learning the uranium business at that time, but he was already well versed and successful when it came to investing in other sectors. Dev had earned an MBA from the University of British Columbia in 1985 and worked for six years as an investment advisor and corporate finance analyst before forming RD Capital Inc. RD Capital is a privately held consulting firm providing venture capital and corporate finance services to emerging companies in the resource and non-resource sectors both in Canada and the U.S.

He'd been President of Lariat Capital Inc. too, which had merged with Medicure, a biotechnology company, in November 1999. Some of his profits from that effort went to keep Strathmore's heart beating. He was also the founder, President and CEO of Royal County Minerals Corp. He sold that company to Lukas Lundin's Canadian Gold Hunter (formerly International Curator) in July 2003, and some of that profit also helped maintain Strathmore.

Keeping the uranium dream alive "wasn't easy, I can assure you," he says, but during another gold show that year, this time in New York, his confidence got a very welcome boost.

"Rick Rule was still positive about uranium and he was joined by others. Bob Quartermain, Jim Dines and Mike Halvorson all said it's time, let's go and we'll support you. I felt vindicated."

Dev already respected Rule's opinions and Quartermain had cast iron credentials as well. Quartermain began his career as an exploration geologist for US Steel and AMAX prior to joining the Teck Group in 1981. By 1985 he'd taken over as President of Silver Standard and spearheaded its phenomenal growth to the point where it controlled the largest in-ground silver resource of any publicly-traded silver company. Like Rule, James Dines was a living legend when it came to accurate investment advice, and Dev had worked with Halvorson for the sale of Royal County. Dev's enthusiasm for uranium was completely renewed by his meetings in New York.

"Rule had said go for it in 1996 and I had, but until that 2003 gold show and meeting with those guys I hadn't received confirmation I was still on the right track. I wanted to be the guy driving the bus but at that time it was like going somewhere without signs. Those guys were the ones giving me the direction I needed."

Dev will always be grateful for the support he received from those industry leaders. From 1998 through to 2003 his dedication to uranium had been viewed doubtfully by others in the investment community. "I was considered the 'village uranium idiot'. I remember guys saying 'there's the uranium guy' when no one else was interested. But I believed in it."

That same fall, Dave Miller rejoined Dev at Strathmore. He'd been teaching uranium geology in China. "We got aggressive and raised money. Very early in 2004, Sprott Asset Management took a position and stared buying into the company and that's when everything started to take off," Dev remembers.

"Our strategy was two-fold. We wanted to buy drilled out deposits because we could get them for pennies at that time, and we wanted to be sure we didn't ignore the blue sky in the Athabasca basin." The wisdom of that strategy may soon be proven to history.

"We got in the uranium business before a lot of new competitors got started up," says Miller. "Being in there early we were able to pick up a lot of great property. Our goals were to acquire known uranium deposits found in the last uranium boom in the 1970s and to have a position in the number one uranium district on earth -- the Athabasca Basin in Canada," Miller recalls.

"We'd already decided our goal when we set out was to put properties into production, so we tried to pick up properties that were the last virgin areas that never went into production," Dev adds.

"Kerr McGee Nuclear Company was the big enchilada back in those days," says Miller. "They saw the writing on the wall back in the 1980s when uranium prices pretty much collapsed and they exited from the sector. They sold their assets to Rio Algom and went back to their core business which was oil and gas." Rio Algom continued operating or maintaining all their properties in New Mexico and even acquired more property in Wyoming, but when Rio Algom was being purchased by BHP Billiton in 2000, the enthusiasm for uranium vanished because BHP management wanted Rio Algom to divest itself of uranium, Miller explains.

"You work hard in life but you still need to have some luck," says Dev. "Mine was the fact that Dave was teaching ISR mining to the Chinese government. It was one of his fellow teachers that told Dave of Rio Algom's plan to dispose of their uranium assets."

On one of the long flights to his teaching engagement there, a traveling companion told Dave he knew Rio Algom was selling its New Mexico assets and the property sounded perfect for Strathmore's strategy.

"Their virgin Church Rock mine and their virgin Roco Honda mine were two properties they had continuously maintained but had never done mining on. They'd done the mine planning but the uranium price collapsed before they went into production. The Roco Honda was going to be the next mine that went online for the hungry 7000-tonne per day Ambrosia Lake Mill they'd built. It was going to be big," Miller says. The Church Rock property is located on the western side of the Grants Mineral Belt.

Historical production in the Church Rock and nearby Crownpoint area was in excess of 16 million pounds. Previous operators had determined the feasibility and amenability of in-situ extraction of the Church Rock deposits. Based on over 150 drill holes, prior operators previously reported over 6 million pounds of contained U3O8 beneath the Church Rock property. A 43-101 technical report for Church Rock estimated 11.8 million pounds of contained uranium and another 3.5 million pounds of inferred resource.

To understand the impact on Strathmore after Dave's traveling companion told him about Rio Algom, Dev offers an example. Imagine if uranium prices collapsed tomorrow and suddenly old mines in the Athabasca Basin came available at fire sale prices. "Where would you go? Cigar Lake? McArthur River? It was the same for us in the U.S." when we heard about Rio Algom's decision.

"It doesn't make a lot of sense to go out and get raw pasture if you have a mine sitting there. Roco Honda was a mine that had somewhere between 15 million and 30 million pounds already. For $150,000 we got it, the entire data base that Kerr McGee Nuclear had in the United States and Church Rock," Dev says happily.

"We're not doing exploration in New Mexico because we don't have to," adds Miller. "We have drilled out deposits. This is where we differentiate ourselves from the other companies. We don't have to go out and find a deposit, so we don't have the risk of exploration in the U.S. We are permitting on our known deposits."

Roco Honda is a bit of an embarrassment of riches for Strathmore Minerals.

Located in the Grants Uranium Belt of New Mexico, between Ambrosia Lake and Mt. Taylor uranium deposits, the Roca Honda uranium property consists of 63 unpatented mining claims located on land owned by the Federal government and administered by the US Bureau of Land Management (BLM). In addition, the property includes one State of New Mexico Lease. The Grant's Mineral Belt yielded over 300 million pounds of cumulative uranium production in its history and was the largest producing uranium district in the world during the last uranium cycle. The combined land package is 745 hectares (1,840 acres) in size. The Roca Honda property is not adjacent to, or near land owned by the Navajo Nation. Based on available drill hole information, a measured and indicated mineral resource of 17,512,000 lbs. U3O8 contained within 3,782,000 tonnes at an average grade of 0.23% U3O8 was estimated for the Roca Honda property. An additional 15,832,000 pounds at an average grade of 0.17% U3O8 are estimated as an inferred mineral resource.

Roco Honda's potential has not been missed by major companies in other parts of the world either. In August, Strathmore inked a joint venture deal with Sumitomo Corp. For a 40 percent interest in the Roco Honda property that Dev bought in January 2004 for $150,000. Sumitomo agreed to invest $50 million into developing the property. That means that Roca Honda is worth $125 million, says Dev, and accidental delineation shows the resource to be larger than expected too.

For example, in September 2007, Strathmore was installing its second monitor well in a four well program for the Roco Honda project when significant uranium mineralization was hit. High grade uranium mineralization, including a 9-foot interval grading of 0.56% eU3O8 (11.2 lbs/tonne) was discovered.

John DeJoia, Strathmore's Vice President of Technical Affairs, said he was not surprised at the discovery. "The results validate our evaluation of the great potential to expand the Roca Honda resource. We expect additional drilling to further demonstrate both the continuity and high quality of this project. Although we have hundreds of drill holes on this property, the team is particularly excited by the fact that this borehole is located in an area that has not been previously explored."

Matt Hartmann, Strathmore's Senior Development Geologist responsible for the Roca Honda drilling program, says the results "indicate the possibility of a new, high grade mineralized trend over a distance of one-half mile in unexplored ground." He added he was optimistic that further drilling will generate additional positive results.

After the Roco Honda acquisition, Strathmore also acquired its 5,000 acre Nose Rock property northeast of Crownpoint and within the Grants Mineral Belt. Its Dalton Pass project of 640 acres also acquired afterward, lies between the Church Rock and Crownpoint uranium districts. With the additional claims staked in the Crownpoint-Dalton Pass areas to complement existing projects, Strathmore's total land holdings in New Mexico now exceed 6,880 hectares (17,000 acres).

But while the New Mexico projects are attracting major interest, it is Strathmore's advanced Gas Hills project, 45 miles east of Riverton, Wyoming, that will likely reach production first. David Miller now manages the company's U.S. exploration and development from an office in Riverton. The Gas Hills project represents a near surface level uranium resource that can be mined by open pit methods bringing with it relatively low technical requirements and environmental risk. There, 100 million pounds of uranium were historically produced.

Miller says he hopes Strathmore can bring on new uranium production in Wyoming by 2010, but the process is still in early stages. Just last February 2007, Strathmore began advancing its Sky and Gas Hills uranium projects to the mine permit application stage. Since, April 2007, the Company has joint ventured 6 non-core exploration and development properties.

Other Wyoming properties include Pine Tree-Reno Creek, Red Creek and the PRB Deposits.

Strathmore has also acquired 1,500 acres in the Wind River Basin as its Copper Mountain property too. Previous drilling defined that there was enough uranium to warrant consideration as a stand-alone mine and heap-leach or conventional mill complex. Originally discovered by Utah International, it was extensively explored by Rocky Mountain Energy in the 1970's and 1980's. Publicly available information puts the Copper Mountain uranium resource size at 20 to 30 million pounds.

The Sky (Cedar Rim) property is another Wyoming project in the permitting process. Sky was originally discovered by Exxon Minerals in the 1970's and was controlled by the Pathfinder group until the late 1980's. Drilling has indicated uranium mineralization of sufficient size to warrant consideration as a satellite in-situ extraction operation. The Cedar Rim property lies northeast of the Gas Hills uranium district.

Strathmore has also acquired property in South Dakota. The Chord property, located in Fall River County, consists of 22 unpatented lode mining claims covering 440 total acres. Previous operators included Union Carbide and Tennessee Valley Authority (TVA). Based on significant exploration drilling an indicated resource of 3.8 million pounds U3O8 was defined, with an additional potential of 11.8 million pounds U3O8 inferred for a total resource of 15.6 million pounds U3O8.

"We're doing our exploration in Canada through Fission Energy Corp. for production 15 to 20 yrs from now," says Dev. "They say business is 80 percent defensive and 20 percent offensive," he adds, explaining the strategy.

"The Athabasca Basin is the second part of our strategy to have some 'blue sky properties' which give us a future," says Dev. "In 2006 we were often approached by larger uranium companies regarding mergers and acquisitions but the discussion was always about the U.S. assets. So, we thought: 'Lets make this discussion less confusing by taking out the

Canadian assets.' The 'blue sky' went to a Fission Energy Corp. (TSX-V: FIS) and all the bigger assets stayed in Strathmore."

In July 2007, Strathmore officially spun-off its Canadian uranium exploration properties along with cash of $500,000.

Strathmore's experience in Canada was similar is some ways to what happened in New Mexico and Wyoming. "In 2004, there was a six- to eight-month window where we were well positioned financially and able to buy a lot of properties."

At the same time as making the Roco Honda acquisition, he says his Canadian team was researching acquisition opportunities in Quebec, Alberta and Saskatchewan. They uncovered a property called Dieter Lake in Quebec that had 25 million lbs. resource. Dev says he got a call from the geologist involved who said "we can get it for almost nothing; in fact I can put it on my credit card." Dev has to chuckle.

"We picked up 25 million pounds. That's the advantage of being in the right spot at the right time. If you put a $4 per pound value on that, you're talking about a $100 million asset we managed to get on a credit card."

"The Athabasca was in pretty good shape and there were still properties to be had," Dev recalls. "Even in 2004, good opportunities still existed so we staked as much as we could. We couldn't believe a big company like Cameco would leave all this land around it free, and we picked it up for very little money."

As a result of the spin-off, Fission Energy has one of the largest portfolios of exploration properties in the Athabasca Basin. Its primary projects in the basin include Waterbury Lake, which surrounds the Areva-Denison Midwest deposit and Davy Lake, where a 51 kilometre long conductor was identified during the 2006 field season. Davy Lake is the largest contiguous exploration block in the Athabasca Basin.

Specifically, at Dieter Lake in Quebec, a NI 43-101 inferred resource totaling 24 million lbs. grading .067% U3O8 has been identified. Significant exploration potential exists to expand this resource, according to Dev.

"The Dieter Lake property," he says, "has a documented history of exploration of about 30 years, though it was considered sporadic with the majority of significant activity occurring in the late 1970's and early 1980's by Uranerz Exploration and Mining. Once again, it represents a significant asset acquired very inexpensively.

"We've executed our strategy pretty much perfectly on the acquisition side," says Miller. "We've never over promoted. When we've been asked how long it would take to get into production we've always said six years from when we start permitting. We are two years into that now and we still think it will take the full six years to get it done so we've still got four years to go. There were many, many other companies that came after us claiming they could permit in two years. They should be in production now but they aren't. In fact, most of them probably aren't as far along as we are."

Miller believes Strathmore will be the first of the new generation of companies to produce a million pounds in the U.S. "Others may get their permits before us but we'll be right behind and frankly our production rate, and our confidence in what we can produce, is pretty high."

Miller gives Dev the credit for getting Strathmore reorganized when uranium prices started to rebound. "He had the vision to be a little bit ahead of the curve and started raising money again. He knew I was familiar with a lot of deposits. We put the money he was able to raise together with the knowledge I had of all these old deposits and we created this space of Strathmore."

"In 2003, I was working out of my dining room table at home. In 2004, I hired one geologist full time and 2005 we hired a couple more and a couple of engineers. Now we've bought an office in Wyoming and have about a dozen people working out of it full time plus a number of consultants. In New Mexico, we have another office that is our permitting and government affairs office. It has about six full time people and probably in excess of 20 consultants working for us. All together we have about 40 to 50 people moving forward on dozens and dozens of properties in the U.S."

Miller says the company began serious staff additions in early 2005 when it established a full-time mine development office in Santa Fe, New Mexico, and hired two professionals, with over 65 years combined experience, to facilitate the mine development process. The mine permitting initiative at the Company's Church Rock, New Mexico property, the Roco Honda project and other efforts, needed a presence.

John DeJoia became the vice president of technical services for the New Mexico operation in the company's Santa Fe offices. DeJoia is a graduate of the University of Wyoming and a Registered Geologist in the State of Wyoming. He has over 30 years of technical experience that includes underground, open pit and in-situ uranium mining. His Wyoming mining experience includes the Shirley Basin and Big Eagle uranium mines with Utah International, Development Geologist for Pathfinder Exploration Corporation, Chief Geologist for Federal American Partners in the Gas Hills District, and Director of Technical Services for American Nuclear Corporation. His extensive management experience includes work for Morrison-Knudsen, Inc. at the Idaho National Engineering Laboratory and Manager of the Washington Group projects at Los Alamos National Laboratory. DeJoia's diversified experience covers the entire range of the uranium mining and milling cycle including: feasibility studies and start-up of new mines; permitting, mining and mapping of ore bodies in new mines; closure evaluations for uranium mine and mill tailings facilities and mixed waste facilities.

Working with DeJoia is Juan Velasquez, vice president of Environmental and Regulatory Affairs. Velasquez is a graduate of the University of New Mexico with an undergraduate Bachelor of Science degree in biology and an MBA. He has over 30 years of experience in the uranium industry including seven years with Phillips Uranium Corporation, a subsidiary of Phillips Petroleum Company (now Conoco Phillips) as Manager of Environmental, Health, and Safety Affairs. He had 15 years experience with United Nuclear Corporation as President of the Minerals Division and Corporate Manager of Environmental Affairs. Velasquez has consulted to private, federal and state clients in the nuclear remediation industry and has permitted several major uranium mining and milling operations, including Phillips Nose Rock mine/mill complex and the United Nuclear Church Rock mill tailings disposal facility. He has managed a variety of environmental project developments, operations, and closure activities for Phillips, UNC, and his clients at locations throughout the United States, working closely with federal and state regulatory agencies to obtain the required approvals. Velasquez is a past Chairman of the New Mexico Mining Association Uranium Environmental Committee and has been active in various other state and national mining associations and organizations.

Other key management personnel include Steven Khan, Strathmore's executive vice president of corporate development. Khan has spent close to 20 years in all aspects of the investment industry, including retail, institutional, corporate finance, capital markets, and investment banking areas. He has held senior management roles in a number of regional and national Canadian investment brokerage houses as well as serving on a number of private and public companies.

Company directors include Michael Halvorson, a successful entrepreneur involved in the securities industry and mining finance since 1967. He is currently a director of a number of public mining and oil and gas companies. He was a director of Viceroy Exploration up until it was purchased by Yamana Gold Inc., and currently sits on the board of Nova Gold Resources Inc.

Dr. Dieter Krewedl, who during a 23 year career with Cogema, was instrumental in the discovery of the Green Mountain uranium deposit in Wyoming, high grade uranium breccia pipe deposits in Arizona and uranium deposits in the Grants, New Mexico mineral belt. Dr. Krewedl presently serves as the President of the Geological Society of Nevada.

Ray Larson is founder of Uranium Resources Inc. (URI) which was one of the few U.S.-based uranium mining companies to survive the industry's extended market downturn. Larson pioneered the exploration, development and production of uranium ore bodies using in-situ recovery (ISR) technology. His experience includes the commercial development of ISR uranium extraction plants at Kingsville Dome and Rosita in south Texas, as well as developing significant uranium mineral interests at Church Rock and Crownpoint in northwestern New Mexico. In addition, he negotiated multiple long-term uranium sale contracts with both U.S. and European utilities, and other industry participants.

What does the future hold for uranium according to this brilliant team?

"There's no immediate, quick solution to the energy shortage out there, but the future is nuclear power and it will be replacing the carbon-based fuels that are depleting rapidly," Miller says. He feels the U.S. is at a turning point where it must act for the future or prepare for new global economic leadership by other countries of the world in the future. "Frankly if we get it (agree to expand nuclear generation solutions as opposed to other energy sources) we will remain an international economic power for the next century or more. If we don't get it, we are going to be a second rate nation within 50 years," he says. As of August 2007, there were 30 reactors under construction worldwide and 439 were in operation. Sixteen of the reactors being built are in developing countries of Asia. According to IAEA statistics, China currently has 26 plants firmly planned and four under construction.

"We have 104 nuclear power plants in this country right now," he says, adding the country needs more and it needs more operating mines. The U.S. generates about 20 percent of its electricity from nuclear power. By 2025, he says electricity consumption in the U.S. is expected to grow by approximately 50 percent from current levels and yet the safe, clean nuclear power plant option is not being widely accepted by all States. Dam projects and coal-fired power plants are still being considered first. And, while demand for nuclear fuel sits at 50 million pounds annually, the U.S. is only producing 3 million pounds per year. "China, 50 years from now, could have many times more nuclear reactors than we do," he predicts, and they will be generating power for an ever expanding economy of tremendous size by buying fuel from the same sources as the U.S. In his opinion, America should be cognizant of the increased demand for uranium fuel around the world and reclaim its place as the largest global producer.

"I'm a believer in the marketplace. The price will go where it has to go to encourage enough new production. The important point is we're consuming twice as much as we're producing without China or India building one more nuclear power plant," says Miller.

"But the fundamentals of the market haven't changed. It's crunch time. People simply can't invest in companies just because they have the word uranium in their name. Investors are going to have to look at the fundamentals and when they do, they are going to like what they see in Strathmore.

"Sumitomo, our joint venture partner in New Mexico, didn't just come to us on a whim. Sumitomo looked at us for over a year. They were already familiar with our projects in New Mexico because these were known projects within the industry. People that know uranium love Strathmore because we have the goods."

Titan Uranium Inc.

leader needs to have vision. And, when it comes to achieving his dream, a leader needs to know the strategic steps necessary to get there. Philip Olson has that instinct.

Olson's father was a miner—first for uranium and then for potash—so it could be said that Philip was born into the business. He grew up at Elliot Lake and played in the historical shadow of the uranium industry. As an adult, Olson became a geologist and spent 30 years ranging across northern Canada in search of mineral wealth. The mines he has worked at throughout his career are as famous for precious metals as his hometown is for uranium. Olson has been Regional Exploration Manager for Hemlo Gold, Chief Geologist at Falconbridge's Kidd Creek, and Exploration Manager/Chief Geologist with Giant Yellowknife Mines. Olson is proud of the part he played in the operation of those companies, but he's now staking his own claim on the future.

"I went to the University of Saskatchewan," Olson says, "and spent my early years as a student and then as a young graduate mapping in northern Saskatchewan for the Saskatchewan government. I worked on a number of map sheets that surround the Athabasca Basin." Like geologists everywhere, Olson gained his mining experience by following the industry's shifting opportunities. When activity in Saskatchewan dried up, he went to Manitoba and worked there for a decade. He went to New Brunswick and across Ontario to Timmins as well. "Then I had the opportunity to come back out to the prairies and I took it. My wife is from Swift Current so I welcomed the chance to come back here with a junior gold producer, Claude Resources." He worked as the Vice-President, Exploration and Corporate Development for Claude until 2005.

"During the eight years I was with Claude Resources, I was also a Director and the Exploration Chairman of the Saskatchewan Mining Association and Co-Chair of the Saskatchewan Mineral Exploration Government Advisory Committee. I used to report on the province's exploration activity, and saw the expenditures by the few companies in Saskatchewan exploring for uranium drop down to their lowest levels as we entered the new millennium. Expenditures were down to just a trickle until the flooding problems that occurred at McArthur River. That event was a trigger point for the uranium price to start climbing. From that I could see an opportunity developing, and in March of that year I acted."

Raising Golden Retrievers in his spare time, Olson's natural bent is that of a hunter—of wild game and of hard-to-find ore deposits. In 2005, Olson decided to pursue the upturning uranium market by building a junior exploration company, Titan Uranium, from a capital pool corporation. Confident in his decades of experience and familiarity with the geology of northern Saskatchewan, Olson is sure Titan Uranium will be as significant to Canadian uranium mining as all those other companies he worked for were to base and precious metals mining.

A scant two years since he began, Titan Uranium has 1.4 million acres in Saskatchewan's Athabasca Basin and a meaningful land position in the prospective Thelon Basin. Titan has in fact become the fifth largest landholder in the world's richest uranium region. Olson has acquired 22 properties hosting hundreds of kilometres of conductors bordering discoveries made by leading uranium majors like Cameco, AREVA, and Denison. In doing so, he's made Titan the only company in the Athabasca Basin with holdings across all six conductive corridors.

Olson is proud of Titan's sustained growth over its first few years of existence, but he is also the first to admit Titan's impressive industry position could not have been achieved without the help of others. Based in Saskatoon, Saskatchewan, Olson's Canadian micro-cap company boasts proven senior leadership and a technical team with over 100 years combined experience in uranium exploration. "What we've managed to accomplish is the result of everybody on the Titan team working together, not just one person's ideas." By focusing on their strong suit as a technical team—acquisition, exploration, and development of uranium properties in the two Basins—Olson believes the Titan group has put their company on the right path and it is ready to flourish. He confidently says Titan is one of the few companies in Canada with the properties, management, and technical depth required to succeed in the cash intensive uranium exploration business.

Titan holds a strategic land position in the Thelon Basin, an area that has had much progress both geologically and politically in recent years. The Thelon, on the border of the Northwest Territories and Nunavut, enjoyed a significant amount of exploration for uranium in the 1970s and 1980s, at least until the price for uranium took a nose-dive and high grade deposits were found further south in the Athabasca Basin. Regarded as a highly prospective analogue of its uranium-rich neighbour in

Saskatchewan, Nunavut's Thelon Basin exhibits similarities in size, age, geology, and mineralization.

During the prior booms a few significant discoveries were made in the Thelon, but the known resources there are still less than ten percent of what has been discovered in the Athabasca. The largest uranium occurrence in the Thelon is the Kiggavik deposit about 80 kilometres west of Baker Lake in the west central part of Nunavut. It hosts an estimated historic resource of 130 million contained pounds U3O8 in a basement-hosted, unconformity-related deposit with mineralization averaging 0.4% to 0.5% U3O8.

In 1978, concerned about the effects mining exploration might have on the caribou herds that calved in their area, the residents of Baker Lake went to court and won an injunction against uranium exploration in their territory. A federal court overruled the injunction however, stating that although the Inuit possessed aboriginal rights to occupy and harvest, those rights did not give the Inuit the legal power to stop uranium prospecting in the Kivalliq (then known as the Keewatin) region. Exploration continued and led to the discovery of several uranium deposits. In the late 1980s a German company, Urangeselleschaft Canada (now part of AREVA Resources Canada) discovered the Kiggavik, Andrew Lake, and End Grid deposits. They proposed to build the Kiggavik mine west of Baker Lake. Opposition to the mine from Inuit hunters and elders led to a municipal plebiscite on March 26, 1990 that rejected the project by 90.2%. Facing such broad opposition, the company asked the federal environmental assessment panel for an "indefinite delay" of the review process for the mine (the present owner, AREVA Resources Canada, is addressing the viability of the project and future plans may include a state-of-the-art mill and mining operation to help alleviate community concerns.)

The Keewatin Regional Land Use Plan, which has been in effect since 2000, prohibits uranium mining in the region until the Nunavut environmental management boards (the Nunavut Planning Commission, Nunavut Impact Review Board, Nunavut Water Board, and Nunavut Wildlife Management Board) have reviewed the mining proposals and the people of the region have approved them. The Inuit are represented in that regard by Nunavut Tunngavik Inc. (NTI), which has adopted a policy in favour of uranium development. Among other reasons, the policy is based on the rationale that encouraging uranium mining and nuclear energy generation will help minimize the impacts of climate change.

"We were the first junior uranium company to make the decision to go back into Nunavut to work," says Olson, describing how Titan anticipated a Thelon Basin uranium renaissance.

"In 2004 and 2005 Nunavut was still stigmatized in the industry by the Baker Lake plebiscite, but we got a sense that the times were changing and that the people were becoming receptive to opportunities. In Nunavut there are 30,000 people and 24,000 are Inuit. Gaining territorial status in 1999, the region has as yet to develop an economic base and I think the government leaders, to their credit, have recognized that mining, and in particular uranium development, must be an integral part of any economic growth strategy for Nunavut. The winds of change were blowing. We saw that. We jumped in. We acquired."

Titan's first acquisitions in the Thelon were 10,000 acres in eight leases. Relying on extensive prospecting and ground survey results from the 1970s and 1980s, Titan Uranium acquired an additional 154,759 acres in seven properties with potential high-grade uranium mineralization, "which is a modest but a reasonable package of land," says Olson. Titan's properties cover about 670 square kilometres on the north-eastern margin of the Basin and are situated about 150 kilometres northwest of Baker Lake and 80 kilometres north of AREVA's Kiggavik deposit. The project area had previously been explored by Westmin Resources with a joint-venture partner CEGB Exploration Canada (now part of Cameco). The joint-venture spent about $5.5 million on programs to delineate. Between 1976 and 1984 Westin had located uranium mineralization in boulders of glacial till with grades ranging from 0.05% to 2.7% U3O8 over narrow widths. Additionally, several boulder trains of unconformity type uranium mineralization had been defined, with grades up to 38% U3O8. "They did all the leg work and developed targets to the drill-ready stage, but then in the early 1980s uranium prices started to crash. Westmin abandoned their uranium search and the properties reverted to a single holder," Olson recalls. Dr. Ron McMillan, a current Titan Board of Directors' advisor, had the properties converted to leases and then maintained them from the early 1980s to the time he vetted them to Titan.

Boasting over 40 years in uranium, gold, and copper exploration and mine evaluation, Ronald McMillan brings an impressive background to Titan. He has served as Exploration Manager for Westmin Resources Ltd. and later as consulting geologist for Cameco Corp., Noranda Exploration Co. Ltd., and Teck Exploration Ltd., among others. Several important gold deposits were discovered under his direction. Graduating from the

University of British Columbia, he has worked in exploration and development of molybdenum and iron ore deposits for Amax Exploration Inc., and as a mine geologist at the Cassiar Asbestos Mine. He then completed post-graduate studies at the University of Western Ontario, receiving a Ph.D. in 1972. He has authored numerous papers on uranium exploration and uranium deposits, and has worked in 20 different countries on five continents. He has served as a director or on committees for numerous professional and trade associations including the Canadian Institute of Mining and Metallurgy and the Geological Association of Canada.

In August 2006, Titan began a summer drilling program to test numerous highly prospective uranium anomalies within the eight Thelon mineral leases. Most of the program's drilling was directed at shallow targets to depths of approximately 80 metres. On December 1, 2006, Titan announced results. The company said that significant uranium values were intersected in bedrock and that this warranted additional infill holes. It was positive news from a company that seemed to be geared to deliver nothing but upbeat reports. What happened in the Athabasca Basin fuelled that reputation.

As well as the acquisitions in the Thelon, Olson also acquired seven properties in the Athabasca Basin. "We launched in June 2005 and in July we acquired the Athabasca properties from a company called the Saskatchewan Syndicate, which was a group of prospectors.

"We wanted to diversify geographically and have properties we could work twelve months a year. In Saskatchewan typically you have sufficient road access for a twelve-month work season whereas in Nunavut you have a very restricted summer season. It is impractical in the early stages of exploration to work anything more than the summer months simply because you are helicopter supported."

The Titan Athabasca Basin Project included seven properties (approximately 310,155 acres) all named after chess pieces. The Castle properties were really the centrepiece to the properties they had. They included the Castle North and South Properties (73,093 acres) in the southwest portion of the Basin. AREVA Resources' past-producing Cluff Lake Mine is within 5 kilometres of the Castle North claims. During its operational history the Cluff Lake Mine produced over 62.5 million pounds of U3O8 at a grade of 1.3%. The Cluff Lake Mine site is currently being decommissioned as its reserves have been exhausted.

A lucky synchronicity helped seal Titan's good reputation in the Athabasca. About a week after Titan took over ownership of the Castle claims, AREVA Resources Canada and UEX Corporation announced spectacular high-grade uranium intersections from three zones at Shea Creek: the Anne, Collette, and Kiana. The AREVA/UEX properties were bracketed on three sides by the newly acquired Castle claims. "We looked at the trend," Olson explains. "The host structure is the Saskatoon Lake Conductor (SLC), a structurally-complex corridor that hosts the nearby Shea Creek mineralized zones of AREVA/UEX. The Saskatoon Lake Conductor trend crosses our Castle North and South properties and we felt acquiring them was pretty much a no-brainer. This is a property that made sense. If you want to find a deposit the best place to start looking is near existing deposits."

For Titan, this represented the first of many timely moves.

Based on the AREVA/UEX discovery, as well as Castle's proximity to the past-producing Cluff Lake mine, Titan commissioned a deep-penetrating geophysical survey of the region to assess the Castle properties' potential. The airborne survey and follow-up drilling clearly demonstrated that the host structure to the Shea Creek deposits traverse both Castle properties. Drilling on the Castle North and South targeted the existence of a basement conductor. At the depths encountered, this conductor zone defines a corridor approximately 3,000 metres wide. The initial drill phase produced widely-spaced pierce points along the trace of this conductor.

Titan's 2006 drilling program significantly advanced the Company's modeling of the Castle South Property approximately ten kilometres south of the Shea Creek Project that hosts the Anne and Collette deposits. Four holes with two wedge cuts totalling 3379 metres were drilled on the Castle South property. Moving north, this drilling intersected the sandstone-bedrock unconformity at progressively deeper levels. The southernmost hole intersected altered basement at 650 metres. The northernmost hole intersected comparable rock types at a depth of 725 metres.

Drilling of the SLC on the Castle North property intersected the basement unconformity at a depth of 1212 metres and the hole returned low uranium values (0.15 to 12 ppm). Two additional holes on the Castle North property were abandoned when the projected depth to the unconformity exceeded 1500 metres. At depths of greater than one kilometre, mining presents challenges beyond current technical capabilities. Future exploration programs on the Castle North property will focus on the north-easterly claims on the rim of the central uplifted segment of basement rocks within the Carswell meteorite impact crater.

Several pronounced electromagnetic conductors parallel the rim of the central uplift and may represent mineralized pathways.

In early April 2007, AREVA and UEX announced plans for a $100 million exploration shaft to further explore the Anne and Collette deposits. Less than a month later, in May 2007, Titan announced that three drill holes were successfully completed on the Castle South project to test the Saskatoon Lake Conductor. The holes completed in this phase of drilling showed elevated to anomalous values of pathfinder elements such as lead, boron, and uranium. Olson says Titan was encouraged by the elevated and anomalous pathfinder geochemistry. "Titan's Castle South property has five kilometres of favourable strike length on the Saskatoon Lake Conductor. Exploration on this property is a vectoring process that requires persistence and a conviction in the geological model. It's a highly prospective trend and we are seeing good geological indicators."

As Olson and his team built the inventory of "chess pieces" that were Titan's Athabasca holdings, the management team began to realize they needed additional critical mass to grow as a company on the rapid schedule Olson had planned. Management began researching other companies with which Titan might do joint ventures or acquisitions. "Near term, in the classic sense with the major companies, is typically a ten- to twelve-year time line," notes Olson. "But as a junior company that is obviously vulnerable to the market, we attach a sense of urgency to everything we do. We've collapsed that twelve-year timeline into a five-year timeline. We're now in our third year, so it's our expectation we are going to be breaking through with something of significance in the next couple of years," Olson predicts.

A partnership with Dejour Enterprises will help Titan make that breakthrough. "We talked with the people at Dejour and agreed that the sum of the two companies put together was far more significant than the individual companies themselves. We did an acquisition of the Dejour property portfolio in the Athabasca Basin. Dejour controlled just about a million acres at that time and Titan had about 300,000 acres. We got their properties and as importantly, we got their technical staff, so we felt it was a very good marriage putting those elements together. We ended up with one of the strongest technical teams with a dominant land position in the Athabasca, particularly on the west and southwest side." The Dejour Enterprises Athabasca uranium asset comprised 68 claims and four permits totalling 966,969 acres. The Dejour projects included the

Fleming, Sand Hill Lake, Virgin-Trend, Meanwell Lake, R-Seven, Maybelle River, and Gartner Lake.

The Fleming project consists of ten claims for a total of 84,819 acres covering several conductors believed to be caused by basement graphitic horizons. A grab sample taken of a pitchblende veinlet on the property assayed at 3.2% U3O8.

The Sand Hill Lake project consists of 18 claims covering 183,628 acres following the southern margin of the Athabasca Basin for more than 25 kilometres. Historic work at Sand Hill identified graphite in basement rocks and a series of northeast-trending faults, the Dufferin Lake structural zone. This same zone hosts the significant uranium mineralization at Cameco's Centennial zone on the adjoining property, where drilling intersected up to 5.83% U3O8 over 6.4 metres and recent drill results intersected 2.48% U3O8 over 19.2 metres.

The Virgin-Trend project consists of four permits and five claims covering 405,763 acres. Prior exploration on the property has been minimal but of note is a large boron and clay anomaly at the south end. Sampling carried out by Dejour in the summer of 2005 confirmed that the boron and clay alteration extends on the property for 20 kilometres.

The Meanwell Lake project covers several basement conductive zones that trend northeasterly across the claims. Meanwell Lake totals 37,038 acres on three claims.

The R-Seven project, which consists of twelve claims covering 134,749 acres, contains numerous graphitic electromagnetic conductor zones. Drilling done on the property in 2006 intersected structurally disrupted and altered sandstone with elevated uranium and lead values.

Both the Maybelle Lake project (three claims totalling 40,706 acres) and the Gartner Lake project (five claims totalling 59,693 acres) lie in an under-explored region of the Athabasca Basin and are overlain by a thick Palaeozoic and Cretaceous sediment cover that hinders ground exploration. However an airborne electromagnetic survey done over the two properties revealed several promising basement conductor zones.

Immediately following final acceptance of Titan's acquisition of the Dejour assets by the TSX Venture Exchange, Robert Hodgkinson, Chairman and CEO of Dejour and Dr. Lloyd Clark, PhD Geology, Director of Dejour, were appointed to the Titan board. Along with the Dejour exploration squad, the appointments gave Titan a team that was highly experienced in the Athabasca Basin. This inventory of skills was boosted even more by the hiring of new members to the team soon after.

At first, leading the exploration team to manage Titan's substantial holdings were Allan McNutt and Dr. Lloyd Clark. Clark is recognized as one of the world's foremost geologists with over 50 years in mineral exploration. In addition to his numerous awards and distinctions, Clark has received twelve fellowships and research grants and published over 23 scientific papers. For nine years, he served as Exploration Manager and Chief Geologist for Saskatchewan Mining Development Corporation (now Cameco) where he established the Exploration Branch and oversaw a staff of up to 65 geologists. His group developed exploration techniques currently employed by numerous companies for uranium exploration, and his team made the discovery of uranium at McArthur River, which is today the world's largest, most profitable, and highest grade uranium mine. Dr. Clark served for six years as Senior Research Geologist and Head of the Geochemical Research Division at Kennecott Exploration Inc. He also served as a Professor of Geology and Geochemistry at McGill University for ten years following three years as pre- and post-doctoral research fellow at The Carnegie Institute in Washington, D.C. Geophysical Laboratory. His vast knowledge would prove to be a guiding force in Titan's operations.

Allan McNutt brought over 30 years experience when he moved over to Titan from his role as Manager of Mineral Exploration for Dejour. He had served for 14 years as Senior Geologist and District Geologist with Rio Algom Exploration Inc. Before that, he spent seven years with Saskatchewan Mining Development Corporation where, as Joint Venture Geologist, he was involved in the delineation of the Dawn Lake uranium deposits and the discovery of the Cigar Lake deposit.

Other members of the technical team included Keith G. Metcalfe, BSc (Adv.) PEng, PGeo—Senior Geologist. Keith Metcalfe had three decades experience in the mining industry and oilfield consulting in Western Canada. And there was John Dixon, MA, PGeo—Senior Geologist. Dixon had managed significant exploration programs for gold, copper, diamonds, base metals, and uranium since 1980. He had worked in Canada and abroad for companies such as Agnico Eagle, Teck Corp., Western Pacific Mining, the Kasner Group, and Watts, Griffis, McQuat. Prior to joining Titan in 2006, John worked for Saskatchewan Industry and Resources, Mines Branch as Exploration Incentive Geologist.

The strength of the board of directors was just as impressive. Hodgkinson, the Chairman, CEO and founder of Dejour Enterprises gave Titan some immediate financing muscle. As well as rebuilding Dejour from the edge of collapse into a significant junior explorer for uranium,

Hodgkinson had experience in the oil patch. He was also the founder and Chairman of Optima Petroleum, and CEO of Australian Oil Fields, which would later merge to become the $400 million Resolute Energy/Cordero. Hodgkinson had been a vice-president and partner of Canaccord Capital, and an early stage investor and original lease financier in Synenco Energy's Northern Lights Project, which today has 1.5 billion barrels of oil in the reserve category in the Alberta Oil Sands.

John Povhe BComm, CA, is the Chief Financial Officer. Povhe brought an impressive background in finance and the resource sector to Titan. His former positions included Controller of the Mosaic Potash Mine in Colonsay, Saskatchewan, Controller of Prairie Malt Limited, and Operations Controller at Claude Resources Inc., a Saskatchewan based gold mining company.

Then, of course, there is Arni Johannson. As the Founder of Titan Uranium, Johannson contributed 15 years of public market experience to Titan. He is also Chairman of Fortress Capital Corporation, a private investment company, and a Director of Mega Uranium.

A huge asset to Titan's strategic planning was the addition of Josef Spross, with his extensive mining background. Spross played an important role in the development and operation of Cameco Corporation's uranium and gold properties. After managing the Key Lake operation for 15 years, he was appointed Vice President of Uranium Mining in 1993, then Vice President of Mining in 1995. In 1996, he was appointed Executive Vice President of Kumtor Operating Company in the Kyrgyz Republic, managing their transition from development to production. In April 1997, he was appointed Senior Vice President and Chief Operating Officer of Cameco Corporation. Holding a Masters Degree in Mine Engineering from Clausthal-Zellerfeld University in Germany, Spross retired from Cameco in 1999 to assume the role of President of the Saskatchewan Mining Association in February 2000 (a four-year term). He currently serves on the boards of Centerra Gold Inc. and RSB Logistic Inc./RSB Services. He became a Director of Claude Resources in 2006.

In April 2007, Brian A. Reilly joined Titan as president. For the 11 prior years, he had worked with AREVA Resources Canada in various capacities in Canada and France, including a two year assignment with AREVA NC, in the parent company's business development group. Brian started with AREVA as District Geologist in the Thelon Basin and worked as a Mine Geologist in the Athabasca Basin before advancing to the position of Director of Corporate Affairs. He served as Vice-President of Human Resources and Industrial Relations and had co-ordinated projects in

Kazakhstan and Nunavut. Reilly brought not only a strong business background to Titan, but also extensive uranium experience in a variety of deposit types, which fit perfectly into Olson's plan for staffing.

Olson also looked to two special advisors to help manage Titan's accelerated growth path — Dr. Ron McMillan and Douglas W. Cannaday. McMillan brought Titan its Thelon Basin opportunities and Cannaday was the President & COO of Dejour Enterprises. Cannaday had three decades of experience having served as director, officer, and controlling principal for a number of exploration companies including Artesian Petroleum Corporation, Amador Resources Ltd., Seahawk Oil & Gas Ltd., and Lava Cap Resources Ltd. In 1999, Mr. Cannaday founded Riomin Resources S.A. in Ecuador. Subsequently, he served as a consultant for several Alberta oil and gas companies before joining Dejour Enterprises in 2003.

"We have 17 people on the staff and 13 of them are geologists. It shows we are technically driven. Perhaps we are myopic to some degree but hopefully we have the right vision. We want to be recognized as a company with a technical focus. Brian Reilly and I can communicate and interact with the marketplace to ensure we are responsive to our shareholder base. They are the people we work for, after all."

"We have $17 million cash in hand. What we've tried to do is put together a very strong technical team with the best properties that are well situated and that are technically viable to explore. We wanted to be able to finance this technical team in the intermediate term so we can fill our five-year mandate to thoroughly explore the ground, and I think we're accomplishing that."

Olson's personal prediction is that "in five years time, I'd like to think we'll be looking back on discovery. I think we are recognized for the elements that we have and that we are a strong company, but we want to emerge as the preeminent junior explorer in Canada."

He says he is not worried about market blips or the industry's future potential at all. "The price climb for uranium has been almost meteoric and we face a cooling off period. My feeling is the price will continue to rise until the cost of electrical generation by nuclear rivals that of coal and even then people will pay a premium on top of that for uranium because it is clean. We will see a second leg up of the price. I think it has a long way to go because it is truly the energy of the future." Olson notes that existing uranium production provides for only 55 percent of global uranium demand. This represents a 70 million pound shortfall in uranium production for 2006, with the balance coming from ex-military sources and shrinking stockpiles. This shortfall is expected to worsen as diminishing petroleum supplies and increased reliance on nuclear energy drives uranium demand from 170 million to 210 million pounds by 2010.

Today, nuclear reactors are responsible for only 16 percent of global electricity, and less than 6 percent of total global energy production. The International Atomic Energy Agency anticipates the construction of over 60 new nuclear power plants in the next 15 years, as well as heightened demand in China and India, which are quadrupling their nuclear capacity by 2020. Given the 15 to 20 years required to bring a uranium mine into full production, and the fact that 66 percent of the world's uranium production comes from just ten mines, the demand on uranium production is expected to continue. Currently, 28 percent of the world's uranium comes from Canada's Athabasca Basin, where Titan Uranium is a principal landholder.

"We think the Thelon Basin may be the next exploration theatre...the new frontier. So we are trying to do deals with companies we think have comparable values." That would include Mega Uranium Ltd., he says. In late June, Olson announced the start of its 2007 summer exploration program in Nunavut and the fact that certain of Titan's Thelon Basin properties were under option to Mega Uranium Ltd. Those properties are located approximately 75 kilometres west of AREVA's 130 million pound U3O8 Kiggavik project. Under the terms of the option agreement, Mega committed to expend $2.5 million in 2007 with an additional $2.5 million in exploration funding required in 2008 to earn a 51 percent interest in the properties. For 2007, Titan was the project manager and operator. "Mega is a very aggressive management group quite prepared to assume the profile of high risk/high reward," adds Olson.

In the Athabasca Basin, Olson has found a similar compatibility with the operating style of Ur-Energy. "We really like the operating philosophy of the Ur-Energy management group and we have done a deal with them on the R-Seven and Rook I properties in the western part of the Athabasca Basin." The two project areas comprise 187,053 acres and are subject to an option agreement with Ur-Energy, whereby the latter can earn up to an undivided 51 percent working interest in the properties by funding $9 million in exploration expenditures managed by Titan over a four year period. The abundance of strong conductors coupled with known uranium occurrences prompt Titan and Ur-Energy to view this project as highly prospective.

Today, Titan is strongly positioned for sustained growth. It has a solid balance sheet, a healthy market capitalization, impressive management and technical teams, and proven prospects. Its next growth spurt will likely come through the systematic evaluation and advancement of its property portfolio. However, Olson says the company is always on the lookout for strategic acquisition or merger opportunities. "We are here to add value the old fashioned way—through discovery. We want to make a difference. Although it may be something of an altruism, I believe our technical team adheres to the philosophy that we all have a debt to society. There is an overwhelming duty to contribute, to give something back. Success in our business comes by way of discovery and that is Titan's driving force."

Tournigan Gold Corporation

Little did the management of Tournigan Gold Corporation know that when they acquired their first gold property in Slovakia, it would ultimately lead them down a path towards holding a number of world-class uranium assets. If the future expectations of the company's president and CEO are met, the story of how Tournigan Gold Corporation claimed a noteworthy place in the European uranium space will be the stuff of legend. As is reflected by its name, Tournigan also has significant gold resources, deftly acquiring a gold prospect in Ireland along with what was once one of Europe's largest past producing gold mines in Slovakia. But it is Tournigan's uranium story that remains the most fascinating piece of its history to date.

At the end of the 1990's Tournigan Ventures Corporation, the predecessor of the current corporate entity, had been an active precious metals explorer. The management at the time had been able to maximize the junior exploration company's promotional cache in Europe and they'd diligently run through the investment capital they'd raised. Unfortunately they found nothing but dust. At the end of the last decade, the company had collapsed with nothing in the coffers. Hopeful shareholders, mostly located in Switzerland, were understandably dismayed when their investments disappeared without the results they'd been anticipating. A group of them attempted to find a legal remedy and Vancouver lawyer Hein Poulus, now a Director of Tournigan, was contacted to initiate some action. Once he investigated, however, he found the Tournigan Ventures principals had been honest and forthright individuals who were basically just unlucky.

Hein had run across a fellow named Damien Reynolds, who was in Australia at the end of 1999. Reynolds was looking to get back in the resources sector and had told Hein he had a dream to build a producing mining company. Hein suggested the failed junior exploration company he'd been investigating could be a candidate for Reynolds' future efforts. Reynolds seized the opportunity and over a period of four years worked his promotional and re-organizational magic with the company and finally emerged with a renamed entity—Tournigan Gold Corporation. In the process of remaking the company, Reynolds had issued 5 million shares of the new Tournigan to Strongbow Exploration Inc. to acquire the

Curraghinalt gold deposit located about 127 km west of Belfast in Central Northern Ireland. At least that way Tournigan had a precious metal asset to match its name. With the deal, Tournigan agreed to issue an additional 5 million shares to Strongbow once a positive production decision was made on Curraghinalt. The property would also be subject to a 2% net smelter royalty payable to Minco plc. "That was a pretty good agreement. No money, just shares," says Jim Walchuck, the current President and CEO of Tournigan.

At that time Tournigan was basically a two-man show — Reynolds was at the helm with a man named Garry Stock as his executive vice president. Stock had worked for Argosy Mining Corp., another junior exploration firm with a gold deposit in Slovakia called Kremnica. As part of Tournigan's reorganization, Stock had asked Argosy's former president and CEO, Yale R. Simpson, to join the Tournigan Board of Directors. "Yale and Damien got to talking and Yale said: 'I know the guys at Argosy really well and they want to get out of gold and into nickel.' Around 2000, gold prices had bottomed out at around $250 an ounce and Argosy had spent a lot of money on this project. They had delineated quite a good resource but there was no investor interest. Their share price was going down and they had found this great project in Africa for nickel that they wanted to pursue instead. Yale suggested Damien offer Argosy some money for the Slovakia project because there was a significant upside to this historically successful deposit." The property, according to Walchuck, was located in one of Europe's largest historical gold mining areas and produced over 1.5 million ounces of gold and 7.7 million ounces of silver since 770 AD.

Reynolds took Simpson's advice, and Tournigan acquired control of what was essentially a one million ounce gold deposit. With the Curraghinalt gold deposit in Ireland and the Kremnica deposit in Slovakia, Tournigan had quickly transformed from a start-up to a junior exploration company with some great potential. "It was fortuitous for Tournigan because at the time no one wanted gold but Reynolds believed in it," says Walchuck.

As he went about completing his acquisitions, Reynolds also started building his management team and a strong board of directors, and asked Dr. Kent Ausburn to join his team as the Vice President of Exploration. Mike Hopley, now a director, was Chief Operating Officer at that time. To establish a strong conduit to contacts in Northern Ireland, he also convinced David Montgomery to join the Tournigan board of directors. Montgomery is currently executive chairman of the Mecom

Group plc., a large European media company with regional offices in Germany, Norway, Denmark, The Netherlands, Poland and the Ukraine. Montgomery was previously chairman of Team Northern Ireland—established by the First and Deputy First Minister's Office of Northern Ireland to stimulate a private sector response to funding major public projects in Northern Ireland.

With men like Hein Poulus, Ronald Shorr, Peter Bojtos, Mike Hopley, David Montgomery and Yale Simpson, by 2004, Tournigan had a strong board and two good assets, "but there was a belief in the market that Reynolds needed more expertise to take the company further," says Walchuck. "His talents were strongly orientated to finance, market savvy and deal making. While he was fantastic at that, it appeared the market thought he needed help to carry out the technical aspects of a future mining company, like scoping, pre-feasibility and feasibility studies, environmental assessments, socio-economic impact studies, and all the rest of the work necessary to move the properties to production. That's where I fit in."

Walchuck joined Tournigan as the Vice President of Mining in March 2004. Walchuck considers himself an underground mining man. During his 28 years in the industry he has worked in Canada, Ghana, and Tanzania. He was the Manager of Mining for Barrick Gold Corporation at the Bulyanhulu Gold Mine in Tanzania where he oversaw the building of a multi-million ounce, high-grade, underground mine in eighteen months. Prior to his six-year stint with Barrick, he was Manager of Mining and then Manager of Corporate Operations for Royal Oak Mines, and before that the Chief Mining Engineer for Tarkwa Goldfields in Ghana.

"Damien put his heart and soul into Tournigan to accomplish the turnaround but it was taking a toll on his family life. So not long after I joined, he asked if I would take on more responsibility, allowing him to spend more time with his family. I became President in May 2005. Then he approached me again in October 2005 and said he'd really prefer if I became CEO and President. He said he would stay on as Executive Chairman and would look after the visionary and investor relations side of things for the company but that I would have full control. That worked out well.

"But in the meantime, Damien couldn't help himself. He started a number of different companies including what I think was ultimately his dream—a company that would act like an investment bank with start-up capital for junior exploration companies. He became so busy by October 2006 that he had to leave Tournigan completely."

Walchuck remembers how in 2004, while Tournigan was attending gold conferences, "the only thing we heard about was uranium." It was more than a bit discouraging. The subject of uranium prospects, who was looking for it and what the predictions were for the rising spot price was the buzz on the floor of every gold conference. "Naturally uranium piqued our interest because we learned the country that held his largest gold assets was also one which produced uranium in the past," recalls Walchuck.

"We phoned our people in Slovakia, particularly our Country Director now, Dr. Boris Bartalsky. Dr. Bartalsky used to look after the geology department of the Geographic Department of Slovakia and he was still very well respected in the mining circles of the Slovakia government. According to Walchuck, Dr. Bartalsky met with the man who had replaced him at the Geologic Survey and discovered that following the collapse of the Soviet Union, the Survey staff had re-assessed all the geology, prospects, and deposits in their country and ranked them from best to worst. We moved as fast as we could to acquire the top three deposits in Slovakia." Matter-of-factly he explains how in doing so Tournigan may have started the wheels in motion for a uranium legend. "As the demand for uranium discovery in the industry was building, we were truly fortunate in having someone like Boris on our team to guide us to some overlooked gems in Slovakia. Upon finding these, we immediately made a strategic investment to do some work on the deposits. This soon led to our acquiring the licences that held about 48 million pounds of historical resource." 1

It is not common knowledge in North America that Slovakia is the third largest user of nuclear energy in the world on a per capita basis, with 58 percent of its energy derived from nuclear reactors with uranium fuel sources from Russia. Until the demise of the Soviet communist empire in 1989, all uranium exploration and mining in Czechoslovakia was conducted by State-owned organizations. In 1993 the state of Czechoslovakia split into the Czech and Slovak Republics. Work on state-funded projects in East-Central Slovakia stopped in 1996 when the former socialist country returned to a market economy system and the government was unable to continue funding its mining projects.

Prior to the collapse of the Soviet Union, the person that was in charge of the uranium exploration in Slovakia—Dr. Laco Novotny—always had secret service watching over him. His team started getting some high grade uranium mineralization hits around a Kuriskova hill in 1985 and about the same time the geologists also made a large discovery in the

Czech Republic that was very economic. The Communist government basically told Dr. Novotny and his people to stand down on all of the exploration and mining of uranium in the Slovak side of Czechoslovakia except for Kuriskova. They continued their delineation of the historic uranium and molybdenum resources. Between 1985 and 1992 they drilled a total of 53 holes comprising 17,936 metres and completed radiometric surveys of each hole. Metallurgical studies carried on during that exploration phase reportedly determined that recoveries of 88 percent to 90 percent uranium and 93% molybdenum could be achieved by pressure leaching of the milled ore with lime and soda at about 150o Celsius.

Dr. Novotny continued the work at Kuriskova until 1992 when exploration funding dried up and he lost his job. "He now works for us," says Walchuck, "and he's having a great time. He's a brilliant man, and nearly two decades after the collapse he still finds it strange that he can talk to North Americans about Slovakia's uranium resources as they were so secretive under communist times. He's also pleased with the technology we're bringing, because it is making a huge difference to him. When Novotny and his team put their uranium resource models together they did not have access to Western technology. Now we dump the data in numeric modeling packages run by high-end computers."

When Tournigan acquired the rights to the Kuriskova deposit, the company also obtained exploration licences for Novoveska Huta. This second property had been an operating uranium mine in the 1970s and 1980s. It was mothballed in 1985 by the Czechoslovakian government when they decided to develop a larger deposit in what would become the Czech Republic. About a year after the first acquisitions, Tournigan added the Kluknava licence to its property portfolio. Tournigan's deposits in Eastern Slovakia are now Novoveska Huta, Kuriskova, Svabovce, and Spissky Stiavnik. Together they cover 98 square kilometres. The 32 square kilometres of exploration claims surrounding the Kuriskova deposit are situated approximately 13 kilometres northwest of the city of Košice, a regional industrial centre in East-Central Slovakia. The project is easily accessible, lying close to the main road between Košice in the southeast and the town of Spišska Nová Ves in the northwest. In addition, the Kluknava licence covers 28.15 square kilometres, including approximately 9 kilometres of strike length of prospective Kuriskova-Novoveska Huta Trend stratigraphy.

After Tournigan acquired the Kuriskova deposit, the company immediately began to update the historic resource estimates to Canadian Institute of Mining, Metallurgy, and Petroleum (CIM) standards. And, following 1,365 metres of confirmation drilling to validate historical drill results, Tournigan engaged London-based ACA Howe International to complete an independent NI 43-101 resource estimate and technical report.

The Kuriskova deposit is blind, covered by thick soils and extensive forest and, as such, this discovery was no easy feat for Dr. Novotny and his team. It has a northwest-southeast strike, and a variably steep-moderate southwest dip. The overall dimensions of the main zone of the deposit start about 120 metres below the surface and extend to a depth of approximately 650 metres. The deposit has been defined as approximately 650 metres along strike by 530 metres down-dip by 2.5 metres thick. The new inferred2 resource estimate that resulted from the ACA Howe calculations indicated that the Kuriskova resource has more than 35 million pounds of U_3O_8.* (Editor's Note: In a news release dated October 24, 2007 Tournigan announced that it expects changes to its May 2007 resource estimate by ACA Howe of 50.5 million pounds U_3O_8. Using an improved geological model, the restated estimate would likely reduce the resource by approximately 10 percent to 30 percent).

"The Kuriskova deposit is beginning to take shape as one of the world's largest high-grade uranium deposits," says Walchuck. So far, Tournigan has only explored a small area of the exploration licence.

For the Tournigan team, the Novoveska Huta project is no less exciting than Kuriskova, though for different reasons. Located on an exploration licence of 13 square kilometres, Novoveska Huta has had considerable underground development completed on it, exceeding 5,500 metres on five vertical levels. There are two smaller deposits and a nearby known uranium occurrence in addition to the main Novoveska Huta deposit as well. Tournigan has not completed an NI 43-101 resource calculation technical report on this deposit yet but it has reviewed the historical estimates and considers them relevant. "Back in the 1970s and 1980s Novoveska Huta was a producing mine. It is ready to go. It just has to be permitted and de-watered, and have some infrastructure put in. But it is fairly low grade. We believe when we put the accumulated data from surface drill holes and kilometres of drifts with channel samples into our computer model, maybe we'll come up with fewer pounds at a higher grade." Novoveska Huta had a historic resource1 of 12 million tonnes of 0.075 percent U_3O_8 or approximately 20 million pounds of U_3O_8, (0.015 % cut-off) at the time of the mine's closing. Since the historical resource cannot be relied upon, a seven hole drill program was designed to twin a number of the original surface drill holes. If successful, the surface drill hole data will then be used to calculate a resource to CIM standards. The Svabovce and Spissky Stiavnik deposits were also former producers. Together the historic resource on these two properties indicates 8,850,000 pounds at an average 0.219% U3O8.

The infrastructure in Slovakia is exceptionally good, according to Walchuck. "It is as good as you see in North America in most cases. Other than the land lines for telephone which are a bit antiquated, the road systems, water, energy are absolutely first class. They've had a hard time of it since 1992 and they are making quite a comeback." He says government systems are much more formalized in Slovakia than it is in Canada "but we have a terrific group of guys over there. The Slovaks are blessed with an excellent educational system. They are learning our ways and we are learning theirs, and it is finally starting to gel very well. It has taken us a while to understand the cultural differences in Central Europe. There is very little English spoken in Slovakia so they are held ransom by what is being said in the press. There has been a lot of damage done environmentally by the communists. Their projects, factories, mines and other industries did not respect the environment, so they look at foreigners very sceptically and at this point don't trust us anymore than they did the communists. I am beyond confident that they will warm up to us," Walchuck says, "provided we continue to perform in a consistent professional manner."

Because the Slovak Republic is a member of the European Union, Walchuck says permitting processes are familiar to the Tournigan team. "It's a very well laid out permitting procedure and very similar to the process you have in North America. You solicit public consultation and work in tandem with government to get things approved. It is the proper way to do things."

Walchuck recognizes that environmental regulations in Slovakia are as stringent as or more stringent than any place in the world, and is committed to following sustainable practices. "Were trying to take all the right steps in the right order at Kuriskova, and at Kremnica as well. But Kremnica is more advanced. It already has a pre-feasibility done on it and Kuriskova has yet to have a good scoping study. We'll probably do that in 2008. It's good that you know that you have a solid regime in place in Slovakia. That is part of the reason we chose to go where we did. We looked at the political risk and the risks with regard to resources. We also looked at the capability of getting something built there with people and infrastructure. Slovakia is a great place to be."

By 2005 Tournigan Gold Corporation had scooped up 1.25 million ounces of gold assets and 48 million pounds of uranium. Despite the assets gained, however, Tournigan was still a 'have-not' company financially. Walchuck recalls, "We had always been hand-to-mouth, raising $2 million here and $5 million there. We were always doing it at 20 cents or 40 cents and always had to give the investors half warrants each time," he says.

Walchuck was approached by Darren Wallace (now of Cormark Securities Inc.) with a fund-raising proposal in January 2006. "Darren said, 'We know a lot of people who believe in your company and who believe in you and they're offering to raise you a bunch of money.'" Walchuck didn't have to think about the proposal for long. An agreement was reached with a syndicate of underwriters led by Sprott Securities Inc., including Dundee Securities and Canaccord Capital. The underwriters agreed to purchase a private placement of warrants to raise $35 million for Tournigan. The underwriters were granted the option to sell additional Special Warrants as well.

The funds were to be used for capital expenditures at the Kremnica gold deposit and to do a scoping study at Tournigan's premium high-grade uranium project, Kuriskova. "At the time our share price was about $1.50. We raised $45 million at that time with no warrants. We basically banked a lot of money we could use to go forward and put towards our programs so we wouldn't have to worry for a couple of years.

"At about that time I was talking with Joe Ringwald, a mining engineer I'd known for 15 years. He was the technical director of mining at AMEC Americas Limited, arguably one of the largest engineering firms in the world, and managing a team of 30 mining engineering specialists in North and South America." Prior to his time at AMEC Americas, Ringwald was Vice President Project Development for Crew Development Corp., with executive control and responsibility for feasibility, permitting, and EPCM of the Nalunaq Gold Project in Greenland. He also held senior positions with SRK Consulting and Placer Dome Technical Services. Over more than 25 years in construction and mining he worked on numerous development and operating projects in North America, Europe, Africa and the Middle East. "I told him I would offer him more money and less stress if he came to work with me at Tournigan. I kinda fibbed about the stress though! He has really added to Tournigan's technical strength."

Ringwald joined the company in April of 2006 as Tournigan's Vice President of Technical Services. He was followed by Steve Stine who became the company's Vice President of Corporate Development. "Steve went over to Slovakia from June to December to try to push forward the pre-feasibility on Kremnica. However, for personal reasons, Steve had to move back to his family in the U.S. before completing his assignment. Luckily, Mike Mracek came back into my life about then." Mike Mracek had thirty-five years of in-depth engineering and operational experience. As the General Manager for the Bulyanhulu Mine, he coordinated technical, operational and maintenance activities. He also worked in Ghana for Ashanti Goldfields Company Limited and had been General Manager for Royal Oak Mines at sites in Ontario, Newfoundland, and the Northwest Territories for six years. Walchuck had known Mracek since 1991 and had worked for him during a three year period on a gold mine turnaround project in Timmins, Ontario.

"Mike and I became very close. Mike's wife Jeanette and my wife, Wendy, were good friends as well. I always kept a spare office vacant in Vancouver so when she would visit I'd say 'this is Mike's office'. I knew she was tired of living in Tanzania. She'd been out of the country for almost ten years and wanted to be back in Canada. I kept putting pressure on Mike, and eventually he came to visit to see his office. He walked in on January 28th and started working on February 2nd as my Chief Operating Officer. He is probably the best operator I've ever had the privilege of working alongside."

The other members of Walchuck's team are no less notable. Michael J. Hopley, Tournigan's Chairman of the Board, is an exploration geologist with over 30 years experience at companies that include Consolidated Gold Fields Ltd., Gold Fields Mining Corporation, Bema Gold Corporation, and Arizona Star Resources Corp. Hopley has been involved in the discovery and development of several large-scale gold deposits such as the Refugio (6 million ounces) and Cerro Casale (23 million ounces) deposits in Chile.

Peter Bojtos, a man with 30 years of worldwide experience in the mining industry, joined Tournigan's Board as well. Bojtos, a geologist and mining engineer, has visited and evaluated properties in 80 countries and been involved in operations in 30 of them. He has carried out 17 significant corporate acquisitions, mergers, or sales in the industry; participated in the development, building, or reopening of 19 mines; and has been involved in the operation of 24 producing mines. He was the regional exploration manager at Kerr Addison Mines Ltd (Noranda Group) at the time of the staking and discovery of what eventually became Inmet Mining Corporation's Troilus gold-copper mine in Quebec.

"In 2006 we also got a fantastic CFO named Hans Retterath and a wonderful corporate secretary named Jude Fawcett. We have a solid team in place now and the technical capability of a mid-tier producer.

"Part of Tournigan's philosophy is to be opportunistic. To us, basically, if it is profitable, makes sense, and can be brought into production or can be grown, Tournigan will act. Those are the key drivers for Tournigan. In the United States; however we've taken to brownfield exploration approach—looking near previously discovered deposits. We figured that for very little money the risk-to-reward ratio was exceptionally high, so we went for it."

Tournigan's United States hunt for uranium resources also began in 2005. The company entered into an option agreement with Sweetwater River Resources LLC to acquire an 85 percent interest in certain uranium properties held by Sweetwater. The agreement allowed Tournigan the option of acquiring the remaining 15 percent interest in the properties at fair value, which the company committed to in October 2007. Collectively, Tournigan's U.S. uranium assets included a property in the northern Arizona Strip uranium district, six groups of claims in three major uranium districts in Wyoming, and three groups of claims in the Southern Black Hills of South Dakota with a historical resource.

Tournigan identified uranium potential in 29 of the 120 breccia pipe targets on the optioned properties in the northern Arizona Strip uranium district. Especially significant was the discovery of uranium at surface in fourteen of those twenty-nine pipes, ranging from a low of 17 ppm to a high of 1780 ppm.

In mid-August 2007, Tournigan's U.S. plans called for at least 20,000 metres of drilling to take place on the Arizona Strip breccia pipes. Walchuck also had drill permits in Wyoming, where the Sweetwater optioned ground covers 37 kilometres in strike length of potential uranium roll front mineralization in the western half of the Great Divide Basin. Infill drilling of a historical deposit, as well as step-out exploration drilling began there in August as well.

"Our hope right now is that we are going to get some very good high-grade hits out of Arizona," he says. "There are a couple of mills very close by that are looking for mill feed so it is a perfect scenario for us. If we find some high-grade ore in Arizona our permitting would not require the building of a tailings dam or a mill. It would basically mean just trucking the ore to another mill north of where we are and there are two mills in Utah." The Uranium One Shootaring Canyon mill is just 225 kilometres from the property.

The sharp edge in Tournigan's U.S. hunt for gold is an earn-in agreement with Au-Ex Ventures Inc., a US precious metals explorer with gold prospects in Nevada. Walchuck says he is pleased to be working with Richard Burdell and Ron Parratt on the joint venture. Ron Parratt is in the Nevada Hall of Fame for having four gold discoveries to his credit, says Walchuck, and he managed the development of resource to reserve at the Twin Creeks, Mule Canyon, Valmy, and Mesquite gold deposits in Nevada. Cumulatively, those activities resulted in the delineation of over 15 million ounces of gold reserves within a 12-year period and the creation of North America's fifth largest gold mining company, which produced over 850,000 ounces of gold during 1996.

"With the Au-Ex arrangement, we don't have to spend a lot of money. We've got some properties, and if they work there is very little money to spend and the reward possibility is good.

"If you take a look at our team," says Walchuck, "you can see we are mine builders. We've all been involved with the Feasibility work. I've got more experience with the hands-on engineering but Mike Mracek has the operational and Joe Ringwald is more slanted to the mine design and feasibility work. That gives us the total package of skills. We are looking at targeting 2010 or 2011 to get uranium production in place.

"We are anticipating that the price may soften about then. There will be ups and downs in the interim too, but I still think long term uranium is a good commodity to be into. There is no doubt about it. If you want to do anything about global warming you have to consider nuclear. After all, you can't build enough windmills and solar panels to make a dent in what is required by the world. The population is growing and needs are growing. It doesn't take rocket science to understand that to provide the necessary energy you can pollute the world with a lot of carbon or you can go with something really clean. The only viable alternative is nuclear. Everybody is realizing that. They are building more nuclear power plants, but they don't have the uranium fuel required so production has to increase."

Walchuck anticipates changes to the uranium market soon. "I think you're going to see a consolidation and a reckoning over the next little while. I think investors got terribly shocked and hurt by the massive drop off in share prices that occurred this year but the money will come back into uranium because it is a good commodity and there is a long term need for it. The fundamentals are very strong but investors are going to be jaded. When they come back in, they will be looking for people who can produce. They'll be looking at the key management a lot more closely I think.

"We're really more focused on sound geology and engineering over promotion because we believe in the end, that's what will pay off for shareholders. I'm not trying to criticize better promoters. I wish I was a better promoter, but all my life I've been in operations. That's the reality. And so is the fact that we have a very strong mine building team and we have world-class deposits in environments where they can be brought into production. I think investors are going to come back and start looking at people, what the infrastructure is like, and the quality of the assets themselves as we continue to develop them. I think we'll probably get a premium when that happens."

So far Tournigan is putting its money where its mouth is, spending approximately $16 million in the ground since entering Slovakia with an ample amount of exploration, development and community involvement on the horizon.

1 Historical Resources

Estimated by Uranovy prieskum, 1992. The Company has reviewed the above historical resource estimates and views them relevant. The historic exploitation of Slovak deposits in general and metallurgical test records acquired with the Company's deposits in particular suggests the reliability of the historical resource estimates. The Slovak category Z-3 is roughly analogous to the Canadian Institute of Mining, Metallurgy and Petroleum's definitions for Inferred Resources However, the Company has not done the work necessary to verify the classification of the resources and the resources are not categorized by the CIM in their Standards on Mineral Resources and Reserves Definitions. The estimate is not to be treated as current and investors should not rely on it.

2 Inferred Resources

This article uses the term "inferred resources". The Company advises US investors that while this term is recognized and required by Canadian regulations, the SEC does not recognize it. "Inferred resources" have a great amount of uncertainty as to their existence, and great uncertainty as to their economic and legal feasibility. It cannot be assumed that all or any part of an inferred mineral resource will ever be upgraded to a higher category. Under Canadian rules, estimates of inferred mineral resources may not form the basis of feasibility or pre-feasibility studies. US investors are cautioned not to assume that any part or all of an inferred resource exists or is economically or legally mineable respectively. However, the

Company has not done the work necessary to verify the classification of the resources and the resources are not classified according to CIM's Standards on Mineral Resources and Reserves Definitions. Investors are cautioned not to rely upon these estimates.

Trigon Uranium Corp.

During an Okanagan summer temperatures can soar into the sweltering high thirties. When that happens it's no surprise to see Sid Himmel, the president and Chief Executive Officer of Trigon Uranium Corp., arriving at his office wearing khaki pants and an open-necked plaid shirt. He looks every part a mining man and even has the careful shuffle patented by field-weary geologists. But temperature has nothing to do with how Himmel is dressed. His sleeves are rolled up to his elbows because he's eager to tackle the uranium hunt for another day.

The way Himmel does business reflects an interesting dichotomy in his background and training. He does everything with the exactitude of a scientist and the practicality of a results oriented financier. In less than a year since joining Trigon, the affable president has remade his company. In that short time he has managed to fill the desks at this micro-cap junior with some of North America's most experienced uranium exploration and development talent. He has also divested the company of the diamond interests that were its corporate genesis and given Trigon a precise uranium focus. Through acquisitions and joint venture deals made since he came onboard, Himmel and the team of mining geologists, geophysicists, and engineers he's assembled have made Trigon Uranium Corp. one of the largest holders of uranium properties in the state of Utah. Trigon has the promise of solid uranium resources in Colorado and Wyoming too, and the additional prospects of small mining operations that could generate handsome cash flow in the offing.

Himmel has a clear vision of where he wants Trigon to be in the next two months, the next quarter, and at the end of the next year. His clarity is possible because he subscribes to a business model that demands the achievement of asset growth every few months. He's guided by precise timelines in his decision making. Growth like that requires Trigon's exploration and development activity to predictably tick over like a Swiss clock. Himmel is the kind of corporate leader who is capable of separating the 'spiel from the real' for investors in the currently heated uranium market. He faithfully applies scientific logic when it comes to project assessment, and, because he has an ingrained investor mindset honed by years on the money side of the business, he's practical about the amount of risk he invites shareholders to accept.

Trigon is headquartered in Kelowna, across the continent from where Himmel spent most of his working life. Before he took over the helm of Trigon, just weeks into 2007, Himmel had accumulated 17 years experience in Canadian capital markets in the nation's financial heartland. He knows the investment community's peccadillos and hot buttons. While in Central Canada, he'd worked for Toronto Dominion Securities as a vice president and director, and for a time was with Merrill Lynch Canada Ltd. as a Corporate Finance specialist in mining finance. During the latter period he participated in the successful funding of numerous growth and large capitalization companies, raising capital in both the public and private financial markets. He also has equity institutional sales and trading experience gained by heading up the Preferred Share Sales desk of TD Securities for two years and having been an institutional equity sales person at that investment dealer from 1992 to 2000. Before all that Himmel practiced as a tax specialist with a Canadian national auditing firm.

Put simply, Himmel knows how to raise a dollar and because he's a Chartered Accountant he also knows how to keep a watchful eye on it for his investors. Additionally, having a science degree focused on chemistry he earned at the University of Toronto means he's equipped to understand the complexities of finding uranium deposits as well. In fact, as a savvy management package goes, the only thing missing with Himmel might be a degree in geology. For that background he can rely on his Directors and the expertise of Trigon's Uranium Development Team.

"I came from a formal business background," says Himmel. The plaid shirt he wears and the company's general surroundings in Kelowna are the antithesis of that image. Trigon Uranium Corp.'s offices, in a low-rise building in the upscale Mission district of Kelowna, are relaxed and informal. It's evident that things are happening there however. Every available workspace is devoted to some task in evaluating or studying the uranium projects Trigon has parlayed. Even the pool table in the staff lunch room has been seconded as a work surface by a young geologist. Himmel's own desk has several computer monitors and the office opens directly to the central reception area so he's handy for all staff and visitors.

"I had a strong interest in operating a company that had consistent returns with minimal downside risks if any and the possibility for substantial upside," he says, explaining why he made the move from watching investment performance for clients to becoming an active wealth creator for investors. A company with diamond resources like Trigon fit the bill, but evolving the business from diamond discovery to operating a diamond mine in Canada's harsh north country was not only horrendously expensive, it was also time consuming — too time consuming for Himmel. "I wanted to be able to provide shareholders a reasonable return on a regular basis equal to or better than the next best competitor on average, and I think you can do that with uranium," he says.

Himmel's strategy is to focus on advanced-stage exploration and property development, rather than join the stampede of junior companies undertaking grass roots exploration in places like the Athabasca Basin. "That way, we can substantially increase the likelihood of success," he says. "We're focusing on lower cost exploration, and therefore on the southwest United States where the exploration costs are 80 to 90 percent less than they might be somewhere else." Himmel explains that the probability of establishing mineral reserves in the neighbourhood of proven producers is much higher than throwing a dart at a map of the Great White North.

To that logic, Himmel is adding well thought out strategies for property portfolio diversification and growth. "We're in the business of uranium development and not undertaking individual risky ventures. We see ourselves operating a portfolio of valuable assets and we apply careful risk analysis and option valuation in our operating and financial decisions." Himmel starts by leveraging the talents of Trigon's senior managers who have experience in applied geological sciences and finance, and then working with the best data they can obtain. "We aim to know, not guess. After we have valued the properties, we only acquire them if they are available at a reasonable cost." And once Trigon acquires a property, the team looks at the full range of possible exploration and development outcomes. "From that we establish expected milestones and reasonable budgets. We call this Management by Objective, a business model which aims to increase organizational performance by aligning goals and objectives throughout our organization."

Trigon has acquired several meaningful projects using Himmel's business model, including acquiring one of Utah's largest holders of uranium properties, Future Energy LLC. Future Energy had over 150,000 acres of uranium claims and an extensive geological database encompassing the entire Colorado Plateau when Trigon made the acquisition. The Colorado Plateau has produced more than 150 million pounds of Uranium 308 and continues to be one of the major uranium districts in the U.S. Following the acquisition in May 2007, the Future Energy office became Trigon's Moab, Utah office. "When we acquired Future Energy, we were convinced that Trigon's business portfolio would expand to involve the profitable syndication of uranium properties in Utah and Colorado and we were right. In fact, we started the process of syndicating such properties immediately and at the same time we were acquiring a number of uranium mines in the Colorado Plateau." Himmel explains the Trigon strategy for the acquisition this way: "Trigon has numerous diversified strategies for growth in the uranium industry. The strategies include acquiring smaller Colorado Plateau mines in both Utah and Colorado where historically ores were processed at centralized mills. From 1949 to 1982, over 11 million pounds of uranium averaging 4.8 pounds U308 per tonne were mined in the White Canyon district where our new mines are located. In Utah, only the Lisbon Valley district has produced more uranium than the White Canyon district."

By early June, Trigon announced it had acquired the mining rights to six Utah mines in the White Canyon area. In mid-summer the company staked an additional 87 claims in the White Canyon uranium district adjacent to its Blue Lizard, Giveaway, and Yankee Girl mines. The recent staking brings Trigon's total up to 160 mining claims, and three Utah State Leases, for a total of about 4312 acres. The properties host six former operating mines. All the mines had been in operation through to the end of the prior uranium cycles in the mid 1980s and, based on U.S. Department of Energy data, the mines produced more than 70,000 tonnes averaging 0.24% U308 per ton prior to 1970.

The Trigon team believe significant uranium resources can be developed by re-examining historical Colorado Plateau mines looking for extentions of previously defined ore zones and exploring land contiguous to the mines. Trigon's Senior Mining Engineer, Thomas Buchholz, began working on mining plans immediately, but the quandary about where to send any ore Trigon's new mines might produce remained a stumbling block. "The uranium industry of the Colorado Plateau is still being defined," says Himmel. Amid those developing terms of reference is the milling capacity in the State. In mid-2007, Denison Mines Inc. had the Blanding Mill and Uranium One Inc. had the Shootaring Canyon Mill and whether either will have time available for the custom milling of Trigon's uranium ore still has to be determined.

In addition to its recent acquisitions, Trigon holds the Marysvale uranium project in central Utah. Acquired in May 2006, Marysvale is a contiguous block of 226 unpatented mining claims and one Utah State mining lease, comprising a total of 4827.5 acres in Sevier and Piute Counties. Located 5 miles east of Marysvale, the property lies in the Basin and Range Physiographic Province at elevations between 6100 and 7200 feet and is accessible by a graded road. Trigon recently completed the first step in a three step exploration program on the property. "The first thing we had to do," Himmel explains, "was gather any historical data and gear up for the first drill program, the second was to drill some test holes, and the third, to begin a definitive drill program." The hope is that the 1.8 million pounds of uranium resources historically estimated to be present in volcanic-hosted, bulk-tonnage, supergene deposits at Marysvale will be greatly expanded by further exploration. "We want to produce things which add to the portfolio, but we want to make sure our data is scientifically accurate. All indications are that we have a chance of 5 to 6 million pounds," Himmel adds, but that has to be proven out to the Trigon President's critically observant satisfaction.

Uranium was first discovered in the Marysvale area in 1949 and production commenced in the district that same year. The deposits were steep, narrow, quartz–fluorite veins hosted within intrusive bodies of quartz monzonite and granite. The veins clustered in a small area that was called the Central Mining Area (CMA) and were mined with conventional track underground methods. A total of 321,000 tonnes of ore were produced from the CMA and milled to produce 1.39 million pounds of yellowcake. In the early 1950s, three claims were staked east of the producing mines in an area which is currently part of Trigon's property. Several adits were driven in near-surface uranium mineralization over the next decade or so. Reportedly high-grade pockets of surface secondary ore were obtained from other pits, trenches, and shallow shafts, but the land was untested by drilling until 1977. At that time Minerals Exploration Company ("Minex"), a subsidiary of Union Oil Company, drilled 24 holes before selling the property to Phillips Uranium Inc., a subsidiary of Phillips Petroleum. From 1978 until 1981—when the decline in uranium prices caused the cessation of exploration activities at Marysvale—200 exploratory holes were punched by Phillips. Much of that data was lost, although collar surveys and downhole gamma logs still exist.

In the mid-1980s a professor of economic geology at Brigham Young University in Provo, Utah, reported on the Marysvale site. At that time, Dr. P. Proctor discussed the nature of the subsurface mineralization there and estimated a blanket-type supergene uranium resource totalling at 1.8 million pounds of U308 existed, with the potential for it to be two or three times that size. His recommendations for follow-up drilling were never carried out before Phillips abandoned the site. The Marysvale site remained in virtual exploration limbo until 2005 when activity in the district resumed with the resurgence of uranium prices. In March 2006,

Garfield Resources staked 100 mining claims in the area and acquired a mining lease on Utah State Section 2, T27S R3W. Trigon eventually aquired the claims and mining lease from Garfield, adding to them by staking 122 additonal claims, effectively consolidating the eastern part of the district. Trigon's land position is now about 30 times the size of the CMA.

Historic uranium production in the district came from steep, narrow veins of magmatic-hydrothermal origin, but the Phillips drilling on the Marysvale property seemed to indicate that relatively flat-lying lenses of supergene mineralization, averaging 30 feet thick, lie near a redox boundary about 300 to 500 feet below the surface. If the Phillips numbers hold true, the potential exists for a much cheaper open pit mining strategy on the property. To add to the historic data, the uranium team, led by Senior Technical Advisor Ian Thompson, initiated a program of geological mapping, radiometric surveys, prospecting, rock sampling, and geophysical surveys on the property. Their program identified several anomalous zones indicative of uranium mineralization. The team found that SP, IP, and magnetic surveys responded well to the known mineralization at Marysvale so geophysical surveys were expanded to cover the rest of Trigon's holdings. The geophysical surveys revealed an area three times the size of the mineralized area at Marysville that demonstrated a comparable geophysical signature. The results begged for some drilling to test the new target zones identified and to confirm the historical resource at Marysvale.

Himmel personally scoured the old Phillips gamma logs to understand what the data revealed about Marysvale. Trigon decided the prospects looked good, but the uranium team still wanted to proceed cautiously. They decided to twin two of the Phillips holes to see if the new results matched the historical numbers and to see if they could extend a highly mineralized zone Phillips discovered in 1981. Results of the Phase One drilling program confirmed the presence of thick continuous zones of uranium at Marysvale. The average uranium content in the mineralized intersections was 0.063% elemental U308, or 1.26 pounds of elemental U308 per tonne using a cut-off grade of 0.02% elemental U308. "It supports our opinion that the property hosts a bulk-mineable uranium deposit," Himmel states. The official report on the results of the 16-hole Phase One program indicates that a major fault zone in close proximity to the main mineralization is an important ore controlling factor in the project area. This information will greatly assist the company in planning the next phases of work at Marysvale. Himmel estimates the next bout of drilling at Marysvale could cost upwards of $6 million and take up to 24 months to complete. It will include a large reverse circulation drill program, diamond drilling, metallurgical leach tests, and baseline environmental studies.

The uranium team is also eagerly exploring another prospect in the Henry Mountains uranium district, adjacent to Denison Mines Corp.'s Bullfrog and Tony M uranium deposits, which reportedly host more than 20 million pounds of U308. The Henry Mountains region is one of the oldest uranium–vanadium areas of the Colorado Plateau, located in the sparsely populated Canyon Lands section. Deposits of uranium were reportedly mined as early as 1913, but because the area is remote the district initially saw only intermittent small-scale mining. In 1948, when the U.S. Atomic Energy Commission began buying uranium, prospecting, and mining in the area increased. For the 30 years leading up to the end of the last uranium cycle production amounted to 475,000 pounds of U308, mostly from small depsits. Exxon and U.S. Energy began looking deeper in the mid-1970s and discovered several large uranium deposits in the southern end of the district, including the Tony M, Copper Bench, Indian Bench, and Southwest ore bodies. The Tony M mine was developed in the 1980s and a uranium mill in Shootaring Canyon at Ticaboo was erected, but it was mothballed after only four months in 1982 because of declining uranium prices. Denison Mines is currently re-opening the Tony M mine and has plans to develop its Bullfrog project as well.

In September 2006, Trigon acquired two property blocks, Henry North and Henry South, covering approximately 19,200 acres immediately west of the Denison projects. Very limited drill hole information exists for the Henry North block, although drilling by previous operators in the area intersected uranium grading between 0.09% and 0.33% U308 there. The Trigon uranium team wants to determine whether the westerly extensions of the known uranium deposits reach into Trigon's property. It's already known that a northwest-oriented trend extends for more than eight kilometres to the edge of Trigon's property, so a drill program to get some more answers was initiated in July 2007. Phase One of that program consisted of about 17,000 feet of rotary drilling involving at least ten closely spaced holes designed to maximize the likelihood of intersecting uranium along trend. Because the known uranium deposits in the Shootaring Canyon area are large and relatively continuous along elongate trends, the Trigon team considers the potential for additional major discoveries in the area to be excellent.

Trigon also has plans to explore property acquired at Wray Mesa, adjacent to Denison Mines Corp.'s Pandora mine, about 50 kilometres southeast of Trigon's Moab, Utah office. The history of uranium discovery in the La Sal region goes back more than 100 years. Prior to 1880 settlers knew that the local natives obtained a yellow pigment from nearby sources. Prior to World War I intensive mining for vanadium began at La Sal Creek, just east of La Sal. The activity declined after the war but was renewed as war threatened again in 1936. In the early 1940s exploration programs were started and development of uranium deposits was stimulated under the Manhattan project. The next 15 years saw active mining and numerous discoveries in the La Sal Creek area. However, the main mineralized trend at La Sal—the Pandora ore body, close to the eastern end of the La Sal trend—wasn't discovered until the 1960s. During the 1970s and early 1980s the La Sal district remained very active with mines being operated in particular by Union Carbide Corporation and Atlas Minerals. These mines included Beaver Shaft, Mike, Snowball, Hecla, La Sal, and Pandora. Production from the district continued until the uranium price declined in the mid-1980s. In 2006 the Pandora mine was re-opened and is currently the district's only active producer, operated by Denison Mines Corp. Until 1979 the La Sal–La Sal Creek district had produced 6.43 million pounds of U308 at an average grade of 0.32%, the highest average grade of any of the Salt Wash districts in the U.S. Since December 2006, Trigon has acquired about 22,000 acres in the district.

Trigon has joined forces with Ur-Energy to tackle a project in Wyoming as well. "Wyoming hosts the largest reserve base of uranium of any state in the US," Himmel emphasizes, "so we're pleased to expand our exploration and development activities into Wyoming." Trigon can earn an interest in the Hauber Project in Crook County. The property includes five areas identified through pre-1987 drilling by Homestake Mining Company and is located near the former Hauber uranium mine which produced 2.63 million pounds U308 at an average grade of 0.225% between 1958 and 1966. While exploring the area surrounding the Hauber mine, Homestake outlined several million pounds of uranium resources which remain undeveloped. Trigon has some very experienced staff operating that project, Himmel says. "Bob Steele, one of our senior uranium mining engineers, will supervise Trigon's work on the project. Bob is very familiar with the project's potential and past exploration as he oversaw operations of the Hauber mine while with Homestake."

To assist him with this diverse and promising portfolio of exploration properties and small mines, Himmel has the backing of a mining dream team at Trigon. The amount of experience included in that list is staggering. Had the current uranium cycle not occurred when it did, Himmel laments that much of that experience could have been lost. "If you look at the uranium experts out there around the world, of 1000 probably 930 are over 70 years old. If the cycle had waited 10 or 15 years to come back, there would be no knowledge base whatsoever." Trigon's team includes Dr. George Poling, Professor Emeritus and former head of the Mining Department and Mineral Process Engineering at the University of British Columbia, serves as the Board Chairman at Trigon. Dr. Poling is one of Canada's leading experts in mineral processing and an authority in the environmental management of mining operations. That's a critical bit of expertise in the realm of uranium extraction.

Then there is Magnus Haglund, Trigon's COO, who has extensive global exploration experience. He was a principal of Trigon Exploration Finland and Managing Director of Oy Alwima Ltd., a subsidiary of Dia Met Minerals Ltd. Magnus teams up with Ian S. Thompson, the Senior Technical Advisor to Trigon. A consulting geologist, Thompson is president of Derry, Michener, Booth & Wahl Consultants Ltd., and was a recent recipient of the Elvers Medal in Mineral Valuation from the Canadian Institute of Mining and Metallurgy.

The powerful mix of advisory talent on the Board is shored up by some heavy-duty advisory expertise on the uranium development team too. Along with Ian Thompson, Trigon has Thomas M. Buchholz as Senior Mining Engineer. Buchholz contributes 36 years of mining experience and was notably engaged in reclamation activities at the Miracle Uranium Mine. Robert M. Steele offers 40 years experience in forming and evaluating mining ventures. He assembled a large holding of uranium mineral rights for Homestake Mining Co. over two decades, advised utility clients about forming mining ventures, and worked as Senior Mining Engineer for Energy Fuels. A half-century of experience is added to the mix with geolgist Byrd L. Berman. Berman has been involved in nine significant uranium projects around the world. As the Manager of Uranium Operations for Marline Oil Corp. he managed a grassroots uranium exploration program that produced a major mineral deposit in Virginia less than three years from inception.

John Nelson, Bruce Norquist, Stuart Havenstrite, and Lisa Hardy also bring their unique wealth of uranium mining experience to the Trigon team's operation. Nelson has provided reclamation services in the areas of safeguarding abandoned mines and mine waste recovery as a Project Manager for the Colorado Division of Reclamation, Mine and Safety. Norquist, as Chief Mining Engineer, offers two decades of experience. For eight of those years he was engaged in the start-up, operation, and closure of the Cotter Corporation's Schwartzwalder uranium mine in South Africa. He was later the Planning Engineer for Teckminco American Inc. Havenstrite has 50 years of mining experience with management duties at ten different mines, and geologist Hardy is a specialist on metal deposits in North Idaho.

As far as the future goes, Himmel's analytical nature keeps him from being overly optimistic. But he is definitely confident about uranium's place in the world's pantry of energy sources and he thinks he understands investor mentality towards the mineral. "Typically in mining, the time from when you explore to the time you have a mine is ten to twelve years. I don't think investors think twelve years ahead though. In reality, you want to work with investors and you want them to make money over that time frame, but you can't be foolishly unrealistic. I think it's better if you can say you'll have a mine in four years, which is why I'm interested in small mines in the southwest U.S. With a small mine you can actually get production. You don't have to go through all the lengthy permitting required for a large mine. That's the place to start."

As mentioned previously however, Himmel notes the challenge is milling. "Before we can start mining we have to know we can get the ore processed at one of Utah's two mills. The mill capacity there is developing slower than I anticipated." Himmel suspects the slow speed at which custom capacity is being made available stems from concerns about the price of uranium. "The price has stabilized in the $100 per pound range," he says, and the industry is slowly moving into a more mature status. "The price is real, but I think people are worrying about when the price is going down as opposed to how much higher it is going to go," and that is making investment decisions for issues like milling capacity more cautious ones.

He thinks price stability will also have a real and quite apparent affect on the nature of uranium stock investments. "As the price of equities has stabilized, people are looking at which players are going to be producing uranium and, hence, cash flow, and which ones aren't. They are trying to establish which ones will be the winners and the losers." That has been the reason one of Himmel's key portfolio strategies has been acquiring a number of small mines where Trigon could be producing ore within eighteen months.

"We're very keen to have good properties for development which is why the Future Energy acquisition was so important for us. We also have a group of prospectors and business people who have worked in the uranium space mostly in Utah and Colorado over the previous two cycles. You can imagine a few of these people are fairly old, but they know their stuff backwards and forwards. They've got amazing databases, they've got terrific relationships with the prospectors in the area, the claim holders, the miners, and the legal community. That basis of knowledge, experience, and relationships will be what keeps our cost down," he believes.

Himmel believes there are huge opportunities awaiting investors in the uranium space. "There are still a very small number of investors in the world devoting themselves to uranium. It really should be followed by an investor who wants to diversify from the other strong energy stocks," he says. "For Trigon, or companies our size, the investor is basically asking what is going to be happening over the next two months. I have to be producing results every two months. That's why I'm picking up smaller mines in the Colorado Plateau," he says. He suggests that investors working with micro-cap juniors have to be aware of the cycles so they don't miss the growth spurt opportunities juniors offer with their short term asset expansions.

Under Himmel's direction, Trigon is operating on a portfolio approach to business that has four parts. The first is finding advanced exploration projects, then re-entering those with smaller mines which can be put back into production. With cash flow then happening, the third part of the portfolio is joint-ventures with larger companies where Trigon is the operator. "Finally, if I satisfy the growth cycle and the financial sides of the plan, I can then do the huge acquisitions. That is what really produces the major value for investors." He feels his logic is sound and proven. "As we move into an investment phase with more mutual fund ownership, this strategy is the correct one. Mutual fund managers are only comfortable with companies that have a portfolio of assets. They don't like binary results."

Uranium is going to remain hot for a long time to come, he thinks. One signal that's correct is renewed investment activity by petroleum companies, he says. "Big oil knows what is happening. In the last cycle every petroleum company had a uranium subsidiary. We're just in the beginning of this cycle but I think that is going to happen again. Investors who understand energy and who are looking for unusually high returns should study this space. It is almost what I call a no-brainer. Uranium should do well for investors." So should Trigon, he thinks. "We've got a terrific bunch of geologists and developers, and there's me on the business side. Hopefully, with the two together, it will work well for investors in the months ahead."

Uranium One

The two parts that came together in the closing days of December 2005, to create what would transform into Uranium One, have definitely formed a larger whole.

The amalgamation of Aflease Gold and Uranium Resources Limited of South Africa and Canadian Southern Cross Resources Inc. completed at the end of that year, had all the makings for a world-leading uranium producer. Commenting on the Corporation's progress at the time, Neal Froneman, President and Chief Executive Officer understated the obvious. "It was a year of significant progress and accomplishments. The merger of the former Southern Cross and Aflease created a mid-tier uranium and gold exploration and mining company, with an exciting portfolio of exploration and development assets in South Africa, Australia and Canada," he said.

Uranium One's overall performance since then has been notable and has maintained that level of excitement. Because some projects the company owned were production-visible, and the fact that the company had a high margin/low cost business model with cash resources, many analysts considered the amalgamation and company creation was one of the most significant events to impact the industry that year. Two years on, the same analysts have expressed disappointment with the speed of Uranium One's development, but the company has nonetheless built the sort of billion dollar valuation most mining company CEO's only dream about. Events that slowed the company's drive toward its forecast production targets in 2007 have not detracted from the lure Uranium One continues to have for investors long-term either. Though it took a dive in share price and slipped from a profit to a loss in the September quarter, Uranium One still has the potential to impact global production of uranium resources in a huge way and for several decades to come.

Maintaining its operational headquarters in Johannesburg, South Africa, Uranium One enjoys control of resource assets in several major uranium producing regions of the world. For example, its Reitkuil Dominion Uranium Project near Klerksdorp, South Africa, hosts one of the world's largest uranium deposits. Its Honeymoon Project in Australia's only uranium-friendly State, is a fully permitted in-situ leach project waiting to come into its own next year. Uranium One has participated in the U.S. nuclear renaissance by building a resource portfolio in Texas,

Wyoming and Utah with production expected in a few years. It has assets in Canada's Athabasca Basin too. And then there's Kazakhstan, the second largest uranium resource base in the world.

Kazakhstan, with a long history of ISL mining that extends back to 1978, has seen a 22 percent compound annual growth rate in uranium production since 1998. There are stated plans to produce 46.8 million pounds of U3O8 in Kazakhstan by 2010 and Uranium One's Akdala, South Inkai and Kharasan projects are expected to provide much of this incremental production.

Uranium One undertook an aggressive expansion strategy since its 'big bang' creation by including Kazakhstan in its portfolio of assets. Now the main test will be to deliver on projects as promised.

In a letter to shareholders, Neal Froneman stated: "We expect to see a strong pricing environment for U3O8 for many years to come as new primary mine supply is required to meet increasing demand from utilities. As an existing uranium producer with many projects in our development pipeline, I believe our company is well positioned to benefit from this situation and that our shareholders will be rewarded as we execute on our strategic vision.

"In addition to geographic diversification, our company is also diversified by mining method with a mix of conventional and in-situ recovery projects. Our portfolio of assets provides a solid base of low-cost and low-technical risk projects to support our unrivalled U3O8 production growth profile over the next several years.

"Our mining method diversification is an example of our strategy in action. We expect over two-thirds of our uranium production to be from ISR mining techniques by 2012, which will distinguish Uranium One from our peers. The global contribution of uranium production by ISR techniques is approximately one-quarter. The advantages of this type of mining technique over conventional uranium mining techniques include: lower capital and operating costs; exceptional environmental friendliness; shorter construction and permitting time lines; and the fact that ISR mining enables lower grades of uranium ore to be extracted economically. For all its advantages, ISR-suitable ore bodies must meet very specific geological requirements and these tend to be found only in Kazakhstan, the U.S. and Australia. We are either producing, or are in an advanced stage of development of ISR projects in all these countries."

Uranium One was already producing at the Akdala mine in Kazakhstan and at the Dominion Reefs Uranium Mine. "Our other conventional mining assets in the United States are well positioned to take advantage of what we believe will be an exceptional uranium pricing environment over the next several years. As an example, our conventional Shootaring Canyon Uranium Mill, which we plan to refurbish, has important permits already in place that will allow us to bring on production years faster than a greenfields project would."

The company has been developing the Hobson-La Palangana project in Texas too, as well as a portfolio of projects located in Wyoming, Utah, Arizona, Colorado and New Mexico.

Uranium One's five year vision at its creative outset in fact, was to achieve annual uranium production of four million pounds U3O8. At first, Uranium One's charismatic president seemed to pin much of his hopes on the Dominion Project because it had two very important factors running in its favor for success. "Dominion is a brownfields project thanks to Anglo American Corporation's mining of its shallow, near-surface deposit in the 1950s and briefly in the 1980s. So, the ore body is well understood and there is a lot of infrastructure there that we can and will use. And the Dominion Project has a significant gold by-product."

A relatively new CIL gold processing plant is part of the infrastructure Uranium One is utilizing to recover both uranium and gold from the Dominion and Bonanza South ore bodies. "The CIL plant is already processing ore from Bonanza South, but it has considerable capacity for further expansion. We're a uranium company and we're focused on uranium, but the gold is a tremendous secondary advantage. The gold credit lowers the extraction cost of the uranium by 20 percent. The higher the gold prices go the lower the uranium cost. We'd be mining uranium free if the gold price went high enough," he says, but Froneman says internal cost models for Dominion were based on a very low price of gold. "It provides a natural hedge for us if there were any foreign exchange gyrations. It offers us diversity, stability and a cost-lowering mechanism." An added advantage is the fact that Dominion Project is very shallow compared to most other South Africa gold operations, he says, so Uranium One will never need to get into the step function requirements of refrigeration and special air systems to reap the gold value.

The NI 43-101 for the Dominion Project declares compliant resources of over 16.1 million pounds of U3O8 in the indicated category and 146.6 million pounds in the inferred category. "Anglo American estimates had indicated a very large resource and that was before there were any codes existing. We did our due diligence to be NI 43-101 compliant and found in the process of doing that, that there was a very high correlation between their data and our data."

The geological database Uranium One inherited from previous owners of the property included data from some 250 historical boreholes and 46,000 data points. Uranium One supplemented that with further delineation and found additional uranium resources.

Uranium One planned on initial production at the Dominion Project of 2 million pounds of U3O8 per year, ramping to 4 million pounds by 2011. In March 2007, the company announced that the processing of underground uranium ore had started at the Dominion mine. The move followed the successful hot commission of the atmospheric leach circuit at the Dominion mill. Froneman said the commencement of ore processing was in line with targets and the company expected to produce 3.8 million pounds of U308 annually from the first phase of the mine. Froneman also said that the Uranium One team was studying if it could boost that production higher by 2011/12.

The company owns prospecting rights on 57,565 hectares around the Dominion property and is completing feasibility studies to increase indicated resources and build two 500 metre declines and a vertical shaft in order to double production to about 7 million pounds annually. That level of production could be sustained for 30 years, according to the company. The vertical shaft would be sunk to 1,000 metres and a ramp then cut from that point to 1,500 metres underground.

"Based on the time we have spent on the ore body, we're certain we have the same reef on the prospecting part of the property," Robert van Niekerk, the company's executive vice president for Europe and Africa is quoted as saying. "The potential exists to at least double production but this could even be a 15 million pound operation."

Froneman is as excited about the other assets too because in 2007, events began to unfold rapidly for Uranium One and there were also some interesting strategic maneuvers undertaken.

That April, Uranium One acquired the mothballed Shootaring Canyon uranium mill in Utah from US Energy Corp. Uranium One also agreed to buy the nearby town of Ticaboo from US Energy Corp for $2.7 million. The town purchase included a mobile home park, a hotel, a restaurant, a convenience store, and a boat storage and service facility. Why buy the town? Uranium One said it plans to house its workers in the town when it resumes operation of the mill.

In May it was reported that production of ammonium diuranate (ADU) had commenced at the solvent extraction circuit at the Dominion mine. ADU, a slurry that contains uranium, undergoes calcining which results in uranium oxide concentrate (U3O8.) At the time, the company expected that within two months sufficient quantities of ADU would have been produced and then calcined to accomplish delivery of yellowcake to the market.

Perhaps emboldened by the progress being made, in June 2007, Froneman announced that his company planned to increase its market share in uranium supply to 15 percent while there was still a gap between supply and demand in the global market. He said the company could only expect to grow its market share by 0.2 percent when the demand supply gap closes by 2015. The uranium market will only increase by two percent annually after that time, the company estimated.

Uranium One's strategy, according to Froneman, is to grow both internally and externally. Its strategy for external growth is based on the fact that larger companies attract higher valuations due to lower perceived risk and utility buyers prefer diversification in production, which Uranium One is attempting to have in place.

Froneman is steadfast in his belief the long-term uranium price will be significantly higher than current levels. As a result, the company is waiting to conclude uranium sales contracts.

"At the moment there is a standoff between producer suppliers and utility buyers of uranium," Froneman quoted as saying. Robert van Niekerk, executive for Europe and Africa agrees and expects the uranium price to reach $150 per pound in the near term.

Once the supply gap is filled, Froneman says Uranium One plans to concentrate on vertical growth, pursuing opportunities in areas such as uranium enrichment, transport and waste management. Froneman said Uranium One would be looking at different technologies involving slightly higher risks than the norm to put the company ahead of its competitors like Cameco, already building conversion capacity.

By July, all was still well at Uranium One. Feasibility studies were started for phased capacity expansion at the Dominion Reefs mine with initial plans to increase ore processing capacity from 200,000 to 300,000 tonnes per month and ultimately double that to 400,000 tonnes per month. The mine shipped its first tanker of ADU on July 9th. Uranium One told shareholders it expected to achieve the feasibility study production target of 491,000 pounds U3O8 during 2007.

Then things began to change. With a downward revision in its production estimates for 2007, in November Uranium One learned just how nervous investors could react to less than positive news. Investors were already fretting after it appeared that uranium stock prices had peaked in late June. In the previous month, the World Nuclear Association (WNA) Symposium in London, heard that uranium resources supply/demand in the global nuclear fuel market was "more than adequate" to satisfy demand with market forces bringing new projects into production. All uranium companies' shares were affected by the fall in the uranium price from the peak of $138 per pound earlier in 2007 to the $75 per pound level it reached in early October due to sales of U.S. stockpiles.

It was then, in late October, that Uranium One announced there would be a drop from 2.5 million to 2.1 million pounds in its total uranium production for 2007. That lit the fuse on a stock sell-off by itchy investors. The company reported that the production decline was partly due to a drop to 200,000 pounds of uranium from previously estimated output of 491,000 pounds at the Dominion Reefs mine plant and the delayed completion of a sulphuric acid plant in Kazakhstan. The company blamed the slide in estimated South African production on an extended commissioning period of its second autoclave. The company's first autoclave had been commissioned and was operating at design throughput, but Uranium One reported it had lost about 100 days at Dominion due to the commissioning challenges.

A darling mining stock of the Toronto Stock Exchange had already seen a major investment company, RDC, change its rating of Uranium One from a 'sector perform' to an 'underperform' status. Because of that little shake to its credibility, and the production decline announcement, Uranium One endured a hectic trading session with a massive 74.6 million shares changing hands. The shift in shares represented over 20 per cent of Uranium One's issued share capital and saw share prices drop 17 percent.

But Uranium One wasn't the only company to suffer weakened share prices.

A month earlier while Uranium One's prices steadily weakened, Froneman asked for shareholders' patience. He said it was unrealistic for shareholders to expect commissioning at the Dominion Mine would be problem free. However, the announcement about the tardy commissioning of the autoclave just added more bad news on top of other information that investors may not necessarily have liked. Most of Uranium One's production in 2007 was coming from Kazakhstan and the company's

South Inkai uranium processing plant had begun processing on schedule with one big, "but" added to the good news. As the company's third producing mine, South Inkai represented a significant milestone for Uranium One. However, the production targets at South Inkai also had to be adjusted negatively as well due to a temporary shortage of sulphuric acid caused by delays in the completion of a local Kazakhstan copper smelter. The shortage would not only impact South Inkai but also the developing Kharasan project which was expected to begin producing uranium in 2008. The company said that along with its Kazakh joint venture partner, Kazatomprom, it is seeking longer-term solutions to the sulphuric acid shortage.

In Kazakhstan, pre-commercial attributable production for 2007 from the South Inkai uranium mine was expected to be 60,000 pounds of U3O8. Attributable production forecast for the Akdala mine, in which Uranium One has a 70 percent stake, remained at 1.8 million pounds U308 at a cash operating cost of approximately US$10 per pound sold. For 2008, the company expects another 1.8 million pounds from Akdala, but Uranium One revised its production forecast downward to 4.6 million pounds from 7.4 million pounds. However, with the continuing ramp-up of its various mining projects, Uranium One anticipated total attributable production of approximately 8 million pounds U3O8 in 2009 and approximately 11 million pounds of U3O8 in 2010. The 2009 and 2010 production forecasts assumed that the Industrial Production Licences for the South Inkai and Kharasan plants in Kazakhstan by the first half of 2009 and that the acid shortage issue could also be resolved by the second half of that year.

It was some sugar to help the sour tasting news go down, but for investors there was also some lingering unease over changes in legislation in Kazakhstan that gave the government power to amend "subsoil" or mining licence agreements. The shortage of sulphuric acid also had shareholders concerned.

In response, Froneman and executive vice president and Chief Financial Officer, Robin Merrifield, conducted Uranium One's first conference call and webcast for investors and analysts after release of its third quarter 2007 results.

Merrifield admitted he and five other directors, had taken a fact finding trip to Kazakhstan to personally get answers for shareholders. Merrifield said he met with the head of Kazatomprom (basically the Kazakhstan atomic energy commission) and discussed the political issues that were concerning some Uranium One shareholders.

"I was assured that potential changes to subsoil rights were not designed for the uranium industry and will not impact the uranium industry. They were very firm about that." He said Uranium One did not expect any changes to its status relevant to the changing sub soil legislation in Kazakhstan.

He added that Kazatomprom officials believed the sulphuric acid shortage was a near term problem only and would be resolved by the mid-2008. "We aren't waiting for that. We're looking at other possibilities for the short term and the long term and we're so far very encouraged by information on resolving that situation."

Those solutions, according to Neal Froneman, include the potential importation of sulphuric acid and in the longer term becoming part of a consortium of companies which will build a sulphuric acid plant in Kazakhstan.

"Once completed it would be expected to secure our acid supply for our current and developing projects. The acid plant is expected to be completed in 2010 and have an annual production of 500,000 tonnes or so of sulphuric acid per annum."

"Our company is in a very solid position today and we have a bright future ahead," Froneman added. He said the company's five year vision now is to maximize shareholder returns and grow Uranium One to become one of the world's top five producers.

"We do not underestimate the challenges of starting up a number of operations. Bringing on new production in the mining industry is seldom easy. And we've had our challenges as we convert our resources into production pounds but I can assure you we have the right team in place, the capacity in place and we have the experience and expertise to get the job done."

The third quarter of the year tends to be lighter for deliveries, Froneman said, which is why sales during that period in 2007 "were lower than expected as the majority of the deliveries scheduled (for the period) were deferred by customers until the fourth quarter."

"Timing of deliveries is usually at the contracted discretion of customers," he added.

He estimated the build up of inventory at Akdala though had a market value of about $90 million, which he expected would be sold by mid-2008.

In South Africa at Dominion, Froneman reported the pre-commercial production of uranium in 2007 was expected to exceed 200,000 pounds U3O8. Dominion produced 86,800 pounds of U3O8 in the form of ADU. This is the first quarter of significant production and the majority of this product was produced from tailing material, Froneman said. It was in line with the sequenced commissioning process Uranium One designed for Dominion Mine.

With the commissioning of the second autoclave, expected by the end of 2007, Froneman said Dominion would have the plant capacity to treat 200,000 tonnes of uranium ore per month. "We anticipate being in a position of being able to declare commercial production at the operation in the first half of 2008."

"Our sales contract at Dominion for the period 2008 – 2010 represents approximately one quarter of the planned production from the operation. We've been prudent in our contracting strategy, not wanting to contract too much material during the start up phase of Dominion. We expect to end 2007 with 200,000 pounds of uranium in inventory which when combined with the expected production for 2008 will provide ample coverage for our existing sales contracts."

Considering Uranium One's efforts in the United States, Froneman said the company now has "a substantial portfolio of development projects and resources" there. He said he expected the first U.S. production to come from the Hobson insitu recovery facility in Texas. Hobson is being refurbished to a production capacity of over one million pounds U3O8 production a year. The refurbishment of Hobson is expected to be completed well ahead of the La Palangana well field, he said. Uranium One is expecting permitting and development of the La Palangana deposit to be completed by the end of 2008 with the initial production of 35,000 pounds U3O8 in the last quarter of 2008.

"In Wyoming we continue to advance our projects in the Powder River Basin," he says. A milling agreement recently signed allows uranium bearing resins from Uranium One owned and operated ISR projects in Wyoming to be processed. The agreement provides for processing and production of up to 1.4 million pounds per year of dried uranium concentrate. Initial ore is expected to be sourced from the company's Moore Ranch project, followed by the Allemand-Ross, Barge and Ludeman projects.

Uranium One submitted an application to construct and operate an ISR facility at the Moore Ranch project in Wyoming. "This is the first new application for a new uranium recovery facility to be submitted to the NRC in the United States in almost 20 years and another clear indication of the nuclear renaissance underway," Froneman says.

"In line with our aspirations to see the Shootaring Canyon Uranium Mill in operation, we have created a team in the U.S. with conventional mining and conventional prospecting experience."

In Australia, Uranium One expects its Honeymoon uranium project, located in the arid northern reaches of the uranium-producing state of South Australia, to begin production in 2008.

"A redesign of the plant continues and is expected to be completed by the end of 2007. We've also changed contractors at the project and as a result production is expected to commence by the end of the 2008" with an estimated attributable production forecast of approximately 50,000 pounds U3O8 produced in that year. "The scale of the project has not changed," Froneman says. This advanced in situ (ISR) project should ramp up to steady state production of 880,000 pounds U3O8 per year. The planned technical processes for uranium extraction have been confirmed at Honeymoon through the operation of a demonstration plant and an 18-month field leach trial. Honeymoon hosts an Indicated Resource base of 6.5 million pounds U3O8 within 1.2 million tonnes at an average grade of 0.24%.

Uranium was discovered at Honeymoon in 1972. Plans for ISL mining started there in 1979 and a draft Environmental Impact Statement was first prepared in July 1982. In the intervening years, the Honeymoon Project became one of the few undeveloped uranium projects in the world to be fully permitted. The mining license for the property is in effect for 20 years.

The advantage of Honeymoon is the relatively low capital cost and low operating cost of the project. A 2004 engineering study by Ausenco projected construction costs of US $24.5 million and working capital requirements of only US $6.3 million, for a total capital requirement of just US $30.9 million.

Uranium One has an indirect interest in seven exploration licences that are considered to have potential for the discovery of economic uranium deposits in the Kyrgyz Republic as well. The rocks that underlie those Fergana Valley licences, and the Santash licence east of Issyk-Kul Lake contain known uranium occurrences hosted in Cretaceous sediments, according to the company. The sediments are considered to hold the potential for discovering large scale uranium deposits.

Previous exploration during the 1950's in the Kyrgyz Republic did not consider the potential for in situ mineable deposits, and the Cretaceous basins remain relatively unexplored for these types of deposits. Of the total of seven licences, five exploration properties are located in the Fergana Valley of western Kyrgyz Republic. The other two properties are located in eastern Kyrgyz Republic, on the east and west ends of Lake Issyk-Kul.

And finally, Froneman looks at Canada as a land of the future for Uranium One.

"Uranium is a sunrise industry. Bottom line, the future of global energy is uranium and I like to say Uranium One will be a significant part of that because we have resources that are world-scale in size."

Through a 50/50 joint venture with Pitchstone Exploration, Uranium One has interest in five properties near Cameco's McArthur River and Cigar Lake uranium mines. "The lure is that its elephant country and it can be a company-maker if you come across a good intersect. We think we have abundant first class resources there."

Four of the properties – Darby, Waterfound, Moon Lake and Lynx Lake – are wholly owned by the joint venture and the joint venture holds an option to acquire a 75 percent interest in the fifth property (Candle), according to company sources.

A $5 million (100% basis) 2007 exploration program has been approved by the joint venture partners. It is anticipated that a 13,000 metre drill program will be undertaken at Darby-Candle, Waterfound and Moon Lake. An electromagnetic survey is also planned for the Darby-Candle and Lynx Lake properties to test for conductive basement formations.

While some constraints have caused Uranium One to delay production progress by six to nine months, Froneman is adamant that there are no fundamental flaws in the Uranium One plans or strategies. "There is absolutely no change in the quality of the underlying assets. We expect to demonstrate the full production potential over the next several years."

Ur-Energy Inc.

Bill Boberg has no doubts that Ur-Energy Inc. will be the next new uranium producer in Wyoming. When he walks the scrub land at this company's Lost Creek project he sees into the future. He can visualize a mine and processing facility in the shadow of the Granite Mountains to the northeast. He can see the piping and pump units of an environmentally sound ISR project that will indelibly mark Ur-Energy's place in American uranium mining history.

As the President and CEO of a dynamic junior mining company with $86 million in the bank, you'd expect him to be self-assured. But, Boberg's attitude has a character foundation more solid than some outward corporate confidence. After all, he discovered the Moore Ranch uranium deposit and several others in Wyoming's Powder River Basin, so he's familiar with what it takes to solve geological mysteries. He's chalked up 20 years exploring for uranium in the continental US and has recorded an equal amount of time gaining experience on mining ground from South America and Africa. Boberg has investigated, assessed and developed mineral resources in some pretty inhospitable places, so when he stands in Wyoming's Great Divide Basin and says this area will soon be a mine, you believe him.

Ur-Energy was incorporated on March 22, 2004 as a private corporation under the laws of the Province of Ontario and now trades on the Toronto Stock Exchange (symbol URE). In the United States, the company has acquired near-term production uranium properties in Wyoming, a State with four major uranium mining districts which together have produced well in excess of 200 million pounds of uranium from sandstone hosted deposits. Presently in the Wyoming region there are two mining operations and in-situ leach plants in production which are operated by the largest uranium producer in the world. Ur-Energy is about to become another producer if Boberg's vision proves correct.

Ur-Energy controls nine properties in Wyoming with the Lost Creek and Lost Soldier deposits currently at an advanced stage. "We've got a very solid resource base of about 25 million pounds of compliant resources on our two major projects that we're taking to production," says Boberg. "That's drilled out of an historic resource base of almost 90 million pounds."

Boberg says Ur-Energy's corporate objectives are three-fold: to start producing uranium in Wyoming by 2009; to be a low cost producer of uranium and not be impacted by lower uranium prices; and to extend and expand the company's production pipeline through aggressive development and exploration programs in the U.S., exploration programs in Canada, and through strategic acquisitions. "Our work to date demonstrates our strong commitment to meeting those objectives," Boberg says. "We are very encouraged by our development work to date at Lost Creek. We congratulate our superb mining team for maintaining a rigorous schedule to accomplish our goals. We are particularly pleased that our projected costs of production at our Lost Creek project will make us a low cost producer and profitable at spot uranium prices considerably below today's prices."

Ur-Energy has 27 employees at its Denver, Colorado head office and 12 employees at its Casper, Wyoming mining operations office. "We have a highly experienced management and technical staff that has nearly 300 total years of direct uranium experience in all phases from exploration and development through to mining. We have the team in place to develop our projects, to mine them, and to market our resources." He adds that the company acquired an extensive database throughout 2005, and the Ur-Energy geologists and mining engineers have since had the opportunity to examine the tens of thousands of files, drill logs, plan and drill sections, and reports it contains since then. "It's a valuable historic database that enables us to properly understand all our projects and create new projects out of our database itself," says Boberg. "Our database consists of over 24,000 drill holes. It cost the historic operators $80 million to generate this data and if we had to replace it today it would probably cost us over $500 million," he estimates.

As Ur-Energy moves its advanced stage properties towards production, the company finds itself breaking a lot of new ground. For example, as Boberg explains, the Lost Creek project has "an aggressive project production timeline and a big part of the reason for that is that the agencies are not used to dealing with moving projects forward at this time. We've taken it on ourselves to be spending significant time working with the agencies and making sure they are fully involved with our projects. As far as we know we are the first of the junior companies to have received a docket number and a technical assignment control number which was issued to us in September 2006 that enables us to deal directly with the NRC.

"We would be expecting our initial mine at Lost Creek to be producing about 1 million pounds a year. We would be ramping up from late 2009 though 2010 and by late 2010 we would be producing the optimum million pounds per year from the deposit. Lost Creek has been giving us excellent pump test and leach efficiency results which approximates recovery. We expect to be getting leach efficiencies for recovery somewhere in the range of 80 percent. We would then expect to be bringing our second deposit, Lost Soldier, into production about two years later on a similar time frame and building it up to production of about a million pounds a year as well."

Boberg says his company has been approached by a quite a number of companies desiring to buy production in advance. "We've been unwilling to do so until we see that we are further along the track of being able to actually have our production and know what our production rate will be. We've still got a ways to go in getting our permitting, our pre-feasibility, and our feasibility and engineering studies completed so that we know exactly what our costs and timing will be. We are quite comfortable waiting at this point in time before we start entering into a long term contract for any of our production. We will not be selling all of our production on long term contracts anyway. We feel the spot market is going to remain fairly strong for several years yet and we want to remain in a position to take advantage of that as well.

"We view ourselves as a full range exploration and development company. We operate more as a major company than as a junior company and we view our position of fulfilling a strategic production mission in the United States. The current mining production world-wide only fills 60 to70 percent of the demand for uranium and the demand is increasing significantly. We've got the opportunity to get our projects into production and be producing for years into the future.

"These projects will be mined by in-situ recovery mining technology which is a matter of essentially reversing the process by which the deposits were placed in these sandstone aquifers in the first place." All the uranium produced in Wyoming since 1991 has been recovered by the in-situ extraction method. In-situ recovery mining is the process of collecting uranium from a water-saturated, underground ore body in a manner which leaves overlying rock strata and the land surface intact. The process involves the installation of a series of wells through which oxygen is injected into the ground water naturally in the sand containing the uranium deposit. Sometimes the natural water chemistry requires that bicarbonate of soda or carbon dioxide also be added to the ground water to enable the oxygen to bring the uranium into solution. The water containing the dissolved uranium is then pumped back to the surface.

The wells are installed similar to most common water wells—with PVC piping. PVC casing is cemented in place, and then piping similar to that used for irrigation is used to transport the water to the injection wells. Similar piping would take the same water, coming out the production well, when moving it to the ion exchange column. "When you come right down to it, this is basically a water plant," says Boberg. "You are dealing with piping and water and oxygen and bicarbonate of soda. There's not much of anything that is going to cause anybody a problem."

The uranium-bearing solution is then piped to a surface plant where a series of conventional chemical processes extract uranium from the solution. The resulting solution, now barren of uranium, is then refortified with oxygen and re-injected into the ore body. This process continues until uranium levels in the production fluid (pregnant liquor) drop to a point where recovery is no longer economical.

The key to environmental safety of the water table at the project site with ISR is establishing an extensive monitoring program through a system of monitoring wells.

"These surround the well fields. Shallow monitor wells watch over any overlying drinking water aquifers or potential drinking water aquifers. The monitor wells are very close to the well field. The mining process is done by pumping at such a rate so it brings the flow toward the production wells themselves," says Boberg. "This assures the ground water flow is not moving the mining solution away from the production wells. From a mining company's viewpoint, it would be a huge waste if we could not control the fluids. We would have a huge expense in not being able to have the fluids go where we want them to. As a result, we carefully set up the process to make sure the fluids are moving the way we need them to go. The monitor wells assist us in knowing that we have control of the water flow. The monitor wells also assist the state government and the Nuclear Regulatory Commission in assuring that we have our fluid flow under control."

Even with such advanced methods, mine building is an expensive game to play. And Ur-Energy is up to the task. "We completed a financing in May 2007 that brought our total cash in the bank up to about $100 million (USD). About $35 million (USD) of that we have set aside for building our plant. Our budget for 2007 is going to be somewhere in the range of about $18 to $20 million (USD) that we will be expending on advancing our projects to production. It will be a similar budget next year and then in 2009 with the $35 million (USD) that we have for building our plant and advancing our projects takes us well into our production. So with the funds that we currently have in hand, we can advance to the point that we are generating income from our projects," he explains.

"The basic fundamentals of the uranium market and of Ur-Energy remain unchanged. Medium and long term uranium supply continue to be a problem. Given Ur-Energy's production schedule, low capital requirements, projected low cost of production and first-class technical team as well as a secure treasury sufficient to take us well into production, Ur-Energy is uniquely well-positioned in today's market. Our production staff is keeping us on schedule to submit our application to the Nuclear Regulatory Commission in October 2007. Our Lost Creek project continues on the fast track to production in 2009. In addition, our exploration and development projects are aggressively moving forward," Boberg says.

With its second quarter financial results for 2007, Ur-Energy reported to shareholders that management had retired the company's only major debt obligation with a $12.0 million payment to New Frontiers Uranium LLC. Roger Smith, Chief Financial Officer, told shareholders that "the primary objective of the Company's treasury and investment policy is safety of principal. Subject to this objective liquidity is emphasized, which allows management to focus its undivided attention on the development of our projects from a well funded, debt-free position."

"We have a very strong exploration to production pipeline at Ur-Energy," he adds. "We've got a number of projects that we're developing as exploration projects that we will be continuing to work on. These are projects where we've got discoveries. We will be moving them forward and developing them into projects for future production. Then we have our development projects that have had a significant amount of work done on them. We will be doing additional development work on those to have them move into our pipeline so that within two or three years after our Lost Soldier project goes into production we will be able to move additional projects into production. So, we fully expect that we will be mining uranium for decades to come in the State of Wyoming."

"At this point in time the projects – Lost Creek and Lost Soldier – are strictly deposits in the ground," he says. "We have good access into the project area but when we actually start construction we will be building a better road network into the project. There is a power line nearby and we will be extending the lines to our site, preparing the site, and then constructing the plant. A number of the items for the plant have a long lead time for construction. We'll be ordering many of those things and getting them on hand during 2008, so that when we get our permit to mine in 2009 we'll have them ready to put into the plant. We won't be waiting for things that have a long lead time to acquire. Then, in addition to the plant, we have to install the well field. We have the drill holes that were put in to define the deposits at this point in time. In defining the specific well field it's a manner of planning the individual wells that will be put in as production wells and injection wells. That planning will be all done in this period of time and so that once we get the permit to mine those wells will be drilled and installed. That will be going on in conjunction with construction of the plant."

In a time of high drilling equipment demand and staffing uncertainty, Ur-Energy has had to face its share of logistical problems. Boberg says. "We have developed a very good relationship with a drilling company in Wyoming and it has been extremely valuable to us to make sure we have that lined up and operating. Experienced staff is also a very dear commodity in the business right now too, because the uranium sector was dead for about 25 years. I have been fortunate to assemble a very strong team of well experienced staff. My management and technical team has a great depth of direct uranium experience, from exploration through production, but we are also recruiting at universities to get younger people into the company. We do have a number, not as many as I'd like yet, but we have quite a number of younger staff being trained by our more experienced staff to get them up to speed as quickly as possible.

"I'm very pleased with the staff that we've got. I think our staff and management are among the very best in the business right now. You'd have to go to some of the major companies to find a staff that is equivalent to ours," he says.

In addition to Boberg, Ur-Energy's management team includes Jeffrey T. Klenda, Chairman of the Board and Director; Paul Pitman, Vice President Canadian Exploration; James Franklin, Chief Scientist and Director; Harold Backer, Executive Vice President; Wayne Heili, Vice President Mining and Roger Smith, CFO and Vice President of Finance, Administration and IT.

Klenda graduated from the University of Colorado in 1980 and began his career as a stockbroker specializing in venture capital offerings. He is a Certified Financial Planner since 1985 and a member of the International Board of Standards and Practices for Certified Financial Planners. In 1988, he started Klenda Financial Services Inc., an independent financial services company providing investment advisory services to high-end individual and corporate clients as well as providing venture capital to corporations seeking entry to the U.S. securities markets. In the same year he formed Independent Brokers of America Inc., a national marketing organization providing securities and insurance products to independent investment advisors throughout the U.S. Currently, Klenda is President of Security First Financial, a company he founded to provide consultation to individuals and corporations seeking investment management and early stage funding. He is also currently CEO, Chairman, and a Director of Aura Silver Inc.

Pitman is contributing his extensive experience as a uranium geologist in northern Saskatchewan to Ur-Energy. Pitman has over 30 years experience as an exploration geologist having begun his career with Gulf Minerals as a project geologist at the Rabbit Lake, Saskatchewan, uranium discovery in 1969. He went on to work as a Senior Geologist for BP Minerals in the 1970s and 1980s, exploring for uranium across Canada. He and Eric Craigie formed Hornby Bay Resources, a lead explorer in the Hornby Bay Basin for unconformity uranium deposits. Pitman was Vice-President of Patrician Diamonds Inc. (1998-2007) and President and a director of Aura Silver Inc. (since October 2003). He was also the Secretary of Nuinsco Resources Ltd. from September 2003 to February 2005.

Franklin has over 37 years experience as a geologist and is a Fellow of the Royal Society of Canada. Since January 1998, he has been an Adjunct Professor at Queen's University and, since 2001, at Laurentian University. He is a past President of the Geological Association of Canada and of the Society of Economic Geologists. He retired as Chief Geoscientist, Earth Sciences Sector, of the Geological Survey of Canada in 1998. Since that time, he has been a consulting geologist and is currently a director of Phoenix Matachewan Mines Inc. (since September 2000), Patrician Diamonds Inc. (since January 1998), Aura Silver Inc. (since October 2003) and RJK Exploration Ltd. (since July 2001).

Backer has over 38 years experience in the mining industry participating in major exploration programs in the commodities of gold, uranium, copper, and phosphate. He has worked for Kalium Chemicals Ltd., Chevron Resources Company, and as Senior Vice President, Exploration for Goldbelt Resources Ltd. As a Consulting Economic Geologist, he has participated in numerous pre-feasibility mining studies as a team leader and in a management position on projects in North America and in the countries of the former Soviet Union.

Teamed with Backer on the mining side is Wayne Heili. Heili's career spans more than 19 years providing engineering, construction, operations, and technical support in the uranium mining industry. He spent 16 years in various operations level positions with Total Minerals and Cogema Mining (now Areva) at their properties in Wyoming and Texas. He was Operations Manager of Cogema's Wyoming in-situ recovery projects from January 1998 until February 2004. Since then, Heili has acted as a consultant for such companies as High Plains Uranium, Inc., Energy Metals Corporation, and Behre Dolbear. His experience includes conventional and ISR uranium processing facility operations.

"In 2006 we spent over $2.5 million doing the environmental baseline studies, drilling, and pump testing at our Lost Creek and Lost Soldier projects. In 2007, we'll probably be spending $6 million to $6.5 million on them for the engineering feasibility and completing the installation of additional monitoring wells." In 2007, Ur-Energy also has about $2 million for exploration in Canada.

Boberg says Ur-Energy has a number of projects that the company will probably not have the time, manpower, or necessarily the money to properly develop. "We'd like to bring in partners that we feel have the ability to develop those jointly with us so that we would be able to advance them more readily than we would if we just kept them ourselves. This is part of our overall strategy, to be able to keep this working for us by having partners maintain these projects. In the state of Wyoming we have 13 individual projects at this time. In Canada we have six project areas that we control at this point in time. Within the database that we have on Wyoming we probably have the ability to create an additional 50 or more projects. Moving projects forward to production is a real time demand and a real financial demand. We feel there's a lot that we can accomplish by bringing on partners to work with us in this manner."

In Wyoming, that policy has resulted in an agreement with Trigon Uranium Corporation to form the Hauber Project. Ur-Energy will contribute its recently acquired property located in Crook County, Wyoming, to the project. That property consists of 205 unpatented lode mining claims and one state uranium lease totalling approximately 5160 mineral acres. The property is over an area of identified uranium occurrences. Pursuant to the terms of the agreements, Trigon can earn a 50 percent ownership interest in the project if it contributes $1.5 million (USD) in exploration expenditures over three years. Trigon will act as manager of the Project.

Ur-Energy also has an agreement with Target Exploration & Mining Company to form the Bootheel Project. For this project Ur-Energy will contribute its southeast Shirley Basin properties located in Albany County, Wyoming. These properties, covering the Bootheel area and the nearby Buck Point area, consist of 269 unpatented lode mining claims and two state uranium leases totalling approximately 6780 mineral acres. Uranium mineralization was intersected on both properties during the late 1970s and the mineralization has the potential to be recovered by ISR methods. Uranium was discovered in the Shirley Basin in 1955 and production continued until 1992. Although the majority of commercial production in the Shirley Basin was carried out by conventional mining methods, ISR methods were tested on several deposits.

"We are pleased with the addition of Target and Trigon to our strategic alliances," says Boberg. "Forming such ventures allow the Bootheel and Hauber Projects to move forward at a faster pace while we continue to bring our Lost Creek and Lost Soldier properties into production."

Ur-Energy's activity in the Hornby Bay Basin in northern Saskatchewan is another good example of that same partnership strategy.

"We've got projects in the Hornby Bay Basin, the Thelon Basin and at the south end of the Baker Lake Basin. In the Hornby Bay Basin we optioned our Mountain Lake and West Dismal properties to Triex Minerals Corporation," he says. At Mountain Lake, Ur-Energy holds 41 claims covering about 95,242 acres that adjoin eight claims held by Triex. The Triex ground hosts the Mountain Lake Deposit, the largest uranium deposit found to date in the Hornby Bay Basin. Near the west end of Dismal Lakes, Ur-Energy's property comprises two groups of 17 mineral claims totalling approximately 34,400 acres. The claim groups cover part of an historical field of uranium mineralized boulders located 40 kilometres northwest of the Mountain Lake deposit.

Boberg says the deal with Triex "allows for the uranium potential of our two properties to be evaluated by experienced explorers with a strong presence in the Hornby Bay Basin and allows us to focus our Canadian exploration on the Screech, Eyeberry, and Gravel Hill Properties in the Thelon Basin.

"Our primary project in the Thelon is the Screech Lake project. It has all the signatures of a buried uranium deposit such as those in the Athabasca Basin. Everything we've done in the area indicates there is a very strong anomaly there." Boberg says. Screech is comprised of 24 contiguous claims covering a total surface area of approximately 62,000 acres. The

Boomerang Lake Mineral Leases of the Uravan/Cameco Joint Venture adjoin the property on the southwest.

Screech Lake was a focus of uranium exploration activities by UG from 1976–1981. The unique Screech Lake radon anomaly was investigated intensively by geochemical, radiometric, and conventional geophysical methods. The source of it was found to be a radioactive aquifer discharging into the lake and along the southwest shores through fissures. The aquifer waters are characterized by very high radon contents and high uranium contents, with high uranium to thorium ratios. The resultant radioactivity causes the lake waters to have increased conductivity due to ionization. Additionally, soils around Screech Lake were found to be very anomalous in radiogenic helium.

"An odd environmental condition makes the Screech Lake project extra interesting, says Boberg. "An uncompleted 1979 UG drill hole, along with corroborating 2005 ground geophysical evidence, indicate that an undefined area, possibly centered on the west end of Screech Lake, has no permafrost. This is in remarkable contrast to the surrounding region, where permafrost thicknesses of 200 metres are the norm."

Ur-Energy has several other exploration projects in the Thelon Basin. The Gravel Hill property consists of 20 contiguous claims covering a total surface area of 50,617 acres. It covers a southern lobe of the Thelon Formation and encompasses two known, basement-hosted uranium showings as well as several isolated uranium occurrences. A large radioactive anorogenic granite underlies the east part of the property. The Eyeberry property consists of 32 contiguous claims covering a total surface area of 62,435 acres. It encompasses two known uranium showings and several isolated occurrences. This property bounds the Thelon Game Sanctuary.

In June 2007, Ur-Energy teamed with an Athabasca Basin player as well—Titan Uranium Inc.—to earn up to an undivided 51 percent working interest in Titan's R-Seven and Rook I properties by funding $9 million in exploration programs. The programs, to be managed by Titan over a 4-year earn in period, began in July with diamond drilling of the properties. The option agreement calls for annual expenditures of $2 million in each of the first three years with a further $3 million in year four. Vesting of a 25 percent working interest will be at Ur-Energy's election after the expenditure of $4 million in the second year of the agreement. Upon the expenditure of an addition additional $2 million in year three, Ur-Energy will be eligible to vest a further 10 percent working interest. The remaining 16 percent working interest may vest with the expenditure of $3 million in year four. Upon completion of the earn-in phase, Ur-Energy and Titan will proceed as joint vent ventures partners with Ur-Energy becoming project operator.

The R-Seven and Rook I properties include seventeen mineral claims totalling 187,053 acres. The claims are located in the southwestern portion of the Athabasca Basin. The unconformity-type deposits of the Athabasca Basin form the world's large largest storehouse of high-grade uranium resources, accounting for 28 percent of total world uranium production in 2006. Individual deposits grade up to 15 percent and 22 percent at the extraordinary Cigar Lake and McArthur River deposits. Grades average about 2 percent for 30 known Athabasca Basin deposits—more than 4 times the average grade of similar type deposits in other sandstone basins of the world. The R-Seven and Rook I claims cover a prospective magnetic trend that hosts a number of strong electro electro-magnetic conductors identified by airborne and ground geophysical surveys.

"This property deal provides Ur-Energy with its first opportunity to participate in the hunt for high-grade uranium in the Athabasca Basin," says Paul Pit Pitman, Vice President, Canadian Exploration for Ur-Energy. "The property was chosen because of the number of prospective conductors outlined by airborne geophysics, interpreted cross-structures, and the fact that the claims are located along the south edge of the Athabasca Basin indicating drillable depths to the unconformity."

Boberg is confident that Ur-Energy's efforts at exploration and development of properties in its current portfolio, added with the joint-venturing plan, will prove a wise strategy for many years to come. "Demand for uranium has been increasing steadily from 56 million pounds in 1980 to about 170 million pounds in 2004. Several analysts following the uranium market believe that demand could reach a level of over 200 million pounds by the mid-2000s."

In April 2007, Jay Thayer, Vice President of the U.S.-based Nuclear Energy Institute said the U.S. is expected to build 33 new nuclear power stations, in addition to the 103 that currently exist in the country. The U.S. already needs 52 million pounds of uranium annually, of which more than 80 percent is imported. The U.S. produces 2050 tonnes (4 million pounds) of uranium annually, but Thayer points out it is hoped the country can increase production to 5000 tonnes (10 million pounds) by 2014.

While uranium production is global, it is controlled by a relatively small number of companies operating in only a few countries. Less than ten companies supply about 80 percent of the estimated world's production while only eight countries supply 90 percent of production. In fact, just four companies produce almost 60 percent of the worlds' supply with the former Soviet Union accounting for 25 percent alone.

Uranium supply is also highly competitive with the utility companies and operates by making individual and undisclosed contracts. Primary mine production accounts for the majority of traded uranium but secondary supplies from decommissioning of nuclear weapons (HEU) and re-enriched depleted uranium from reactor fuels and uranium tails are significant. Even so, those alternatives to primary production represent less than 10 percent of supply.

Contracts with utilities for purchases of uranium are formed with producers on either a medium (less than 5 years) or long-term basis (more than 5 years) with delivery of the uranium two to three years after the signing of the contract. Pricing formulas are complicated and not made public by either side. Utilities also purchase uranium through spot and near-term purchases from traders as well as producers. The market has moved from the inventory liquidation era into an era of strong demand.

"The only answer to the supply gap appears to be higher mine production," says Boberg. "That's where Ur-Energy comes in."

Waseco Resources Inc.

When meeting the president of a junior mineral exploration company, there are a few things one might assume. For instance, one might expect to be greeted by a map-toting geologist, sitting behind a desk littered with rock samples. Or, alternatively, one may expect to shake the hand of a buttoned-up finance whiz with a degree in business, armed with financial acumen and stock reports. It's likely a safe bet that you would not expect to be seated across the table from a former tennis pro.

But that's exactly who greets you when you are introduced to the affable president of Waseco Resources Inc. In an industry where rock hounds and money-men abound, where expectations drive explorations, Rick Williams is an unexpected and refreshing find. And, upon reflection, it's not so far fetched. The qualities that make for a professional athlete—focus, determination, passion—are the same qualities necessary to thrive in the cyclical mining business. In that respect, Rick Williams is well-equipped. He knows what it takes.

That being said, it takes more than a good backhand to drive a mineral exploration company. Williams has it covered. It just so happens, as he sits down to chat about his company in the late summer of 2007, he is celebrating 30 years at the bar. And it's his legal training that ultimately drew him into the world of mineral exploration, first as corporate in-house counsel and Vice President for mining companies for ten years, and eventually in his current role as President of Waseco, a junior company with some encouraging Canadian uranium prospects.

"I wanted to be a lawyer practicing in Europe," Williams says of his early ambition. "But in order to specialize as a Canadian lawyer in Europe, I thought I better have some Canadian exposure. So I worked here for a year and then had an opportunity to get involved in the junior mining business as in-house counsel to a number of successful companies. I took that on and found it interesting."

Williams worked as in-house counsel from 1980 to 1990 for respected companies like Republic Goldfields and Minefinders. He found himself immersed in buying and selling properties, developing mines, and doing securities work for companies. "One of the things I found is that a number of the mining companies were run by geologists and engineers who didn't really know how to talk to the street," he says. "That was one of the hats that I wore as in-house counsel and one of the things that was needed by a number of the companies—particularly those who didn't have offices in Toronto—I would talk to the analysts and money managers and brokers on their behalf."

Although he was born in Montreal, Williams lived most of his youth overseas in such far-flung locales as Jamaica, Paris, Costa Rica, and St. Louis. He returned to Canada to pursue his law degree. With globetrotting in his veins and fluency in three languages—English, French, Spanish—Williams has found the mining industry to be a perfect niche for his expertise and interests. "Partially because it has allowed me to use my languages and it allowed me to travel. And I was always fascinated by what makes all these juniors go up and down."

Now he's got a front-row seat at the helm of Waseco, a company that has acquired a highly prospective uranium land package. The properties, pragmatically referred to as Blocks I-V, comprise some 700 claims covering 330 square kilometres in the historically significant Quebec-Labrador Trough area. Waseco has completed their own geophysical work, soil sampling, and geochemistry on the property, combined with an extensive compilation of historical work. Williams is confident that they are in a good position and poised for exciting discoveries ahead. Historic reports on Waseco's property show the proven presence of gold, copper, iron ore, and, most importantly, uranium. Some geologists have suggested that the applicable model of uranium mineralization could be the Olympic Dam, the world's largest uranium deposit;

others have drawn comparisons to the uranium rich Athabascan Basin model "While this sounds a little ambitious, the potential for a world class deposit does exist," Williams notes.

Hosting such a promising uranium portfolio, it's no surprise that Waseco was approached by at least four parties interested in a joint venture. Williams chose to go with UraMin, a senior player with a focus entirely on uranium that possesses the necessary technical expertise to advance uranium properties. "There are so many companies looking for uranium, but the looking part is only a portion of what has to be done," Williams notes. "You also have to be able to develop it, bring it to production, and these people have all the skills. They're bringing in uranium mines in Africa right now. That's a very important aspect to what they're doing. They're not just funding, they actually have the technical skills to explore and develop properties."

Under the joint venture deal, UraMin pays Waseco $300,000 and by spending $1.6 million over two years, they can earn a 50 percent interest in the project. By spending the additional cost of taking the project through to a successful feasibility study UraMin can earn an additional 20 percent. "Under the deal we have with them, we retain all non-uranium assets, all non-uranium minerals found on the property," Williams says, adding that if they're combined with uranium—a not uncommon occurrence—it remains part of the joint venture. "But if we find a gold deposit that has purely gold, or if we find diamonds or copper, that would be 100 percent owned by Waseco."

Currently, UraMin is a Toronto-listed $2.5 billion company based out of South Africa and London, which, in itself, is clearly impressive enough. But by the time of the initial fieldwork in Quebec in the fall of 2007, Waseco may find itself with an even bigger partner. "UraMin is just in the process of a friendly takeover bid by Areva, one of the largest energy companies in the world…I think they have 61,000 employees," Williams says.

Things are moving quickly, and in all the right directions, which is the way Williams likes it. Going into only the second year, they have already increased their claims four-fold, joint-ventured with a savvy, responsible partner, spent about $1 million in geophysical work and identified promising targets. Once the targets have been confirmed on the ground, drilling should get underway. "It is quite exciting," Williams says. "This is probably the most exciting part of the exploration phase."

Things may have moved quickly over the past couple of years, driven by the hot uranium market, but it's been ten years since Williams first began his relationship with Waseco. When the company initially came to his attention in 1997, it was a shell that had been acquired by one of his associates. The name Waseco, he explains, originally stood for Washington State Exploration Company. "They were doing some exploration in Washington State before my client acquired it," Williams says. "They looked and didn't find anything and dropped the property, and the name hasn't been changed." When asked if he would be changing the name to something more suitable, Williams was noncommittal. "We haven't made it a priority. When we find something, we might become more demonstratively uranium oriented."

Before uranium took off and caught the interest of Williams, he got down to business looking for a good fit for Waseco. After examining about 80 different possibilities, he found an appealing gold project in Indonesia. "I put together that deal, which went into the shell called

Waseco. We spent about $5 million exploring in Indonesia, developed a significant gold resource, and took it to feasibility study."

Little did anyone know at the time that "gold", "Indonesia", and "1997" would become synonymous with the ill-fated Bre-X fiasco. Waseco's Indonesian gold mine dreams were quickly tainted with the subsequent scandal, as were countless other such ventures. "We waited for a year to see if it would blow over. It didn't, and in the wake of Bre-X we have, to date, been unable to secure the financing to bring it into production." In the meantime, they haven't completely abandoned the Indonesian project which has been put on care and maintenance and is currently in the process of being re-exposed to the market, seeking a new joint venture partner. "There are people who have expressed interest so we have revived our effort on that project," Williams says.

At the time, however, the Waseco team had to set aside their disappointment and seek other opportunities. As Williams points out, they couldn't sit around waiting for the Bre-X stench to clear the air. "We had to consider what to do. We had a company, it was costing us X dollars a year, and we knew we had better do something else."

Among other things, in 2002 and 2003, Waseco took on a diamond project in northern Ontario next to the Victor deposit—a major discovery owned by De Beers in the Attawapiskat area. Because the terms of the earn-in were somewhat "demanding", Waseco decided not to pursue the project. However, they have hung onto the promising property, participating on a 5 percent basis, and they may increase their interest in the future. "There's a world-class mine being developed there, adjoining our property," Williams points out.

Although the gold and diamond projects are still in the wings, most of Waseco's efforts and attention are now focused on their current uranium properties which, ironically, came about because of interest in the Indonesian gold project by a British investment house. It hired mining consultants to review the project and their assessment was enthusiastic about the Indonesian project, but expressed some reservations. "I was told that they said it's a very robust project and someone will make a lot of money with it, however, because of the perceived country risk and the type of deposit—it's an alluvial gold project—the British and North American markets wouldn't like it as much." Because of these concerns, the consultants recommended the deal go through a private company.

"Since the investment house was interested in public sector investments, they then asked if we had any uranium properties and as a matter of fact, on our desk, we had some proposals from some other people and we said, 'If you're interested in them, have a look.'" They were very interested. The only caveat for Williams was that Waseco acquire the properties, which they did. A joint venture was then struck that gave UMC Energy plc an earn-in-right in exchange for project funding.

UMC was so keen to become bigger in uranium their people hurried over to look at operations in the former Soviet Union. While there though, according to Williams, they found oil and gas opportunities that sparked their interest. "They went over there looking for uranium and acquired some oil and gas assets. They then changed their focus to oil and gas and gave us back the properties with the exploration work that we had done together."

With their promising uranium properties back in hand, Waseco was subsequently contacted by UraMin's people who were interested in joint venturing the property and, in the end, the deal was struck. There was never any question of not joint venturing, according to Williams. "Our share price was too low," he said. "Our currency is the properties. That's where we generate the greatest value."

If the promise of their properties is any indication, the value generated should be worthwhile for all involved. Situated about 200 kilometres north of Schefferville, in the Quebec-Labrador Trough, Waseco's extensive property covers approximately 330 square kilometres in five block areas. Two years of geophysical work and compilation of historical data have identified 15 zones of uranium enrichment and 157 discrete uranium anomalies, and the land between Blocks I and II has now been joined into one continuous block. This area has been the primary focus of exploration during the past season and features confirmed multiple large zones of surface uranium.

"There appears to be an iron formation that runs between Block I and II and the airborne geophysics suggests that the uranium is at the base of this iron formation," Williams said of their reason to stake the property in between the two blocks. "So, they have now been joined and we have the whole length of the iron formation, which is about a fifteen kilometre strike length – hopefully with 15 kilometres of uranium at the base of it."

These optimistic scenarios notwithstanding, Waseco and its shareholders can take reassurance in the fact that this area of northwestern Quebec is not an unknown entity. Before the collapse of uranium prices in the 1970s, a number of major mining companies were actively exploring in the area.

The results were encouraging enough that several drill programs were underway before the price of uranium plummeted.

In today's market, with uranium prices skyrocketing from $7 to around $100 a pound during the last four to five years, Waseco is in the fortunate position of holding prospective properties with a proven partner. Uranium is enjoying a great resurgence in popularity and price for many reasons, most notably as a fuel of choice, as well as its application in food preservation and in the field of medicine, where radioisotopes are widely used for diagnosis and research. "They started off hoping this would be a cure for cancer," Williams said. "There's still some research being done in the area."

More importantly, it's the demand for nuclear power that has fueled the race for uranium and the consequent increase in price. Right now, nearly 17 percent of the world's electricity is generated from uranium in nuclear reactors. "There's been a shift, even from the environmentalist point of view, because there are no greenhouse gases, so it's a cleaner fuel," Williams says. "It's more efficient and the risk of nuclear disasters by way of spills or contamination has been reduced significantly over the years." Another element of the equation Williams points out is that the cost of uranium is a very small component of operating a nuclear plant— only about two percent. "So whether it costs $100 or $200 a pound really doesn't factor in to the economics of using nuclear as a source of power."

Whether uranium cures cancer, or offers a cleaner, more environmentally friendly fuel source, there's no question it's become a hot commodity and the number of companies jumping all over it has grown dramatically. "I think there's 600 companies now in the uranium sector with varying qualities and quantities of properties," Williams notes, adding that this compares to about four majors and three or four juniors just a scant three or four years ago.

By the end of 2007, Waseco should have initial results from a field program. Given the size and the scope of the programs, Williams foresees sticking around for "a number of years". Their vision is long-term, and mining is definitely part of the game plan. "We have the ability to mine it because we have UraMin as our partner and together we have the expertise to do that. We've made that decision, and it's a key one, because there's been a 30-year dearth of any work in that field, so experts in uranium are now few and far between.

"We've done very well to team up with one of the best, if not the best, in the industry."

Williams is sharing Waseco's story in late summer 2007, at which point they have an $18 million market cap and a share price of 60 cents, a position that could change quickly.

"Everybody's waiting for us to announce we're in the field and drilling," he says.

Although Williams concedes it's too early to say how much tonnage lies in the ground, he's optimistic. "We're looking at potentially world-class deposits, given the size of the targets that we've identified." The "we" referred to by Williams, includes not only himself and the UraMin partners, but his team in Toronto, where Waseco's head office is located. It's a lean and experienced management team, which is how Williams wants to keep it. He is adamant about maintaining a modest overhead cost and ensuring that funds are used primarily for in-ground exploration. "Very low overhead and no salaries. We're in it for the share appreciation," Williams says. "It's not being done on promotion, it's being done on performance."

One of the key members of the Waseco team is its Chairman, A. C. A. (Peter) Howe. Howe worked in Indonesia and was the founder of A.C.A. Howe International, a world-class recognized geological consulting firm established in 1960. "He's been in the exploration business forever," Williams says "and he knows everybody and all the properties. He was instrumental in selecting the Indonesian properties initially. Howe also has expertise in uranium because A.C.A. Howe was the project manager for one of the largest uranium deposits in Australia," the Jabiluka uranium deposit.

Rounding out the expert management team, Geologist Derek Bartlett brings 35 years of experience to Waseco. He was Vice-President, Exploration for Goldfields and has been active for the last ten years as both a CEO and Director of several juniors. Jay Richardson, CA, is the CFO for Waseco and a number of other mining companies. All share an entrepreneurial spirit and a keen interest in international travel. On the technical side of things, Waseco will be relying primarily on its partner UraMin, but also has a local manager, Tom Sills, who is looking after the field crew and equipment on the project.

"And that's it," Williams says, summing up the Toronto team. "We're stakeholders. The management are stakeholders in this, so the way that we're going to make money is by success in the ground. Our philosophy is to find deposits that can be developed."

With most of the groundwork laid on the projects, it's about to become very active up north for the Waseco team. Waseco's land package is quite remote, situated roughly 200 kilometres northwest of Schefferville, the nearest major centre for transportation. Otherwise, there's no infrastructure near the property aside from a camp a few miles away owned by Uranium Star, which Williams says they have been kind enough to allow us to use "They have a twenty-five-man camp and they just built a runway. Uranium Star has made some very good exploration progress on the ground and enhanced the possibilities within the area."

Among other things, Waseco will be engaging in talks with the local Native bands, which are primarily Cree. Company policy insists on open communication and relations-building within the local communities. According to Waseco's mandate, upholding its social, environmental, moral, and ethical responsibilities is "essential" for both the community and company growth. Williams says he'd like to start discussions with the local Native bands as early as possible. "If we can employ some people and do things for the community, we'd like to do that," he emphasizes. "As soon as the crew is on the ground, it will be one of the discussions our people will be having with the local bands."

Williams acknowledges that it's the process of creating something, creating wealth where none previously existed, that brings satisfaction and drives him. He alludes to another project he's involved in that is currently bringing a mine into production and hiring up to 100 people directly and indirectly. "I feel pretty good about that," he says simply, adding that he's confident Waseco is heading in the same direction.

Whether he's recalling fondly his glory days of playing with tennis greats like Jimmy Connors, or his globetrotting youth lived in various romantic European locales, Williams is earnest and appreciative of the opportunities and experiences he's had. Right now though, the only thing that beats out Waseco and its uranium prospects for Williams' devotion is his family: his wife and two daughters. Fortunately for all of them, Williams' chosen profession allows him the freedom to combine work and family, which he does as often as possible.

"We were in Italy over March break looking at gold properties in Tuscany. We stayed in a 13th century house with metre-thick walls and almost no phones in a little village of 200 people. We spent a week there just touring around. It was very nice." The good news for shareholders is that even when Williams is vacationing, he's working. But his favourite place in the world, the place he calls "home", even though he doesn't live there yet, is Paris. "Every time I get off the plane, I'm at home there," he says wistfully.

When asked if Paris is where he will choose to retire, Williams offers a rare, but engaging grin: "I'm already retired," he says brightly. "This is sort of my second career now."

It looks like this fifty-four-year-old multilingual lawyer, former tennis pro, and current president of Waseco, has no intention of slowing down. He is committed to his projects and his company and currently anticipating the thrill of discovery that awaits everyone concerned in the Quebec-Labrador Trough. With drills ready to hit the ground and the necessary homework done on preferred selected targets, the outlook for Waseco is as positive as it gets in this industry.

"I think we have some properties of merit and we went with the strongest exploration and technical team, we could find," Williams asserts. "I wanted to ensure that we got the best kick at the can."